Euler

The Master of Us All

Typesetting by Integre Technical Publishing Co., Inc.
Cover design by Freedom by Design.
Images of Euler appearing on pages xxvii, 172,
and the cover courtesy of Gerald L. Alexanderson.

Complete Set ISBN 0-88385-300-0
Vol. 22 ISBN 0-88385-328-0

Printed in the United States of America

Current printing (last digit):
10 9 8 7 6 5 4 3 2

The Dolciani Mathematical Expositions

NUMBER TWENTY-TWO

Euler

The Master of Us All

William Dunham

Truman Koehler Professor of Mathematics
Muhlenberg College

Published and Distributed by
THE MATHEMATICAL ASSOCIATION OF AMERICA

THE
DOLCIANI MATHEMATICAL EXPOSITIONS

Published by
THE MATHEMATICAL ASSOCIATION OF AMERICA

––––––––––

The DOLCIANI MATHEMATICAL EXPOSITIONS series of the Mathematical Association of America was established through a generous gift to the Association from Mary P. Dolciani, Professor of Mathematics at Hunter College of the City University of New York. In making the gift, Professor Dolciani, herself an exceptionally talented and successful expositor of mathematics, had the purpose of furthering the ideal of excellence in mathematical exposition.

The Association, for its part, was delighted to accept the gracious gesture initiating the revolving fund for this series from one who has served the Association with distinction, both as a member of the Committee on Publications and as a member of the Board of Governors. It was with genuine pleasure that the Board chose to name the series in her honor.

The books in the series are selected for their lucid expository style and stimulating mathematical content. Typically, they contain an ample supply of exercises, many with accompanying solutions. They are intended to be sufficiently elementary for the undergraduate and even the mathematically inclined high-school student to understand and enjoy, but also to be interesting and sometimes challenging to the more advanced mathematician.

MAA Service Center
P. O. Box 91112
Washington, DC 20090-1112
1-800-331-1622   fax: 1-301-206-9789

For Penny,
who has made all the difference.

Contents

Acknowledgments

During the 1997–98 academic year, I was privileged to hold the Donald B. Hoffman Research Fellowship at Muhlenberg College. For their part in granting me this award, I thank Muhlenberg's Faculty Development and Scholarship Committee, Dean Curtis Dretsch, and President Arthur Taylor. Without such support, I could never have begun—let alone completed—this project.

Likewise, I acknowledge the two libraries whose resources proved essential over the past year: the Trexler Library of Muhlenberg College, where this manuscript was written; and the Fairchild-Martindale Library of Lehigh University, where Euler's *Opera Omnia* can be found in all its glory.

Among those who encouraged this work, foremost is Don Albers, Director of Publications and Programs of the Mathematical Association of America. Don has been a constant source of good advice and good cheer. For his professionalism and his friendship, I shall always be grateful.

Others made significant contributions, most notably Bruce Palka, who edits the Dolciani Mathematical Expositions; Elaine Pedreira and Beverly Ruedi, who shepherded the manuscript through production; and Jerry Alexanderson and Leon Varjian, whose careful reading of the document generated a number of helpful comments.

I also wish to express my appreciation to friends and colleagues in Muhlenberg's Department of Mathematical Sciences: George Benjamin, Roland Dedekind, Margaret Dodson, Linda Luckenbill, John Meyer, David Nelson, Bob Stump, and Bob Wagner.

Beyond the Muhlenberg campus are individuals deserving special mention. One is George Poe, Professor of French at the University of the South, a dear friend without whose translations I would have committed one *faux pas* after another. Likewise, I thank Claramae and Carol Dunham and Ruth and Bob Evans, all of whom have been steadfast in their support. And there are our sons Brendan and Shannon, who remain the greatest.

Finally, and most sincerely, I wish to recognize my wife and colleague Penny Dunham, who made numerous improvements to the final manuscript and who applied her computer wizardry to generate the figures contained herein. For such assistance—and for so much more—I dedicate this book to her with love and thanks.

W. DUNHAM

Allentown, Pennsylvania

"Read Euler, read Euler. He is the master of us all."

–Laplace

Preface

In the crypt beneath St. Paul's Cathedral lies the tomb of Christopher Wren, architect of that great and beautiful building. The accompanying inscription ranks among the most famous of epitaphs:

Lector, si monumentum requiris, circumspice.

This translates as, "Visitor, if you seek his monument, look around." Indeed, an architect could have no finer memorial than the huge church soaring overhead. From nave to dome, from transepts to choir, St. Paul's is Wren's masterpiece.

Mathematics lacks the tactile solidity of architecture. It is intangible, existing not in stone and mortar but in the human imagination. Yet, like architecture, it is real. And, like architecture, it has its masters.

This book is about one of the undisputed geniuses of mathematics, Leonhard Euler. His insight was breathtaking, his vision profound, his influence as significant as that of anyone in history. Euler contributed to long-established branches of mathematics like number theory, analysis, algebra, and geometry. He also ventured into the largely unexplored territory of analytic number theory, graph theory, and differential geometry. In addition, he was his century's foremost *applied* mathematician, as his work in mechanics, optics, and acoustics amply demonstrates. There was hardly an aspect of the subject that escaped Euler's penetrating gaze. As the twentieth-century mathematician André Weil put it, "All his life...he seems to have carried in his head the whole of the mathematics of his day, both pure and applied."[1]

If the quality of his achievement was extraordinary, so too was its sheer quantity. At present, 73 volumes of the *Opera Omnia* (his collected works) are in print—a publishing project that began in 1911—and many volumes of scientific correspondence and other manuscripts are yet to appear. Euler was

---

[1] André Weil, *Number Theory: An Approach through History*, Birkhäuser, Boston, 1984, p. 284.

a veritable Niagara, one who wrote mathematics faster than most people can absorb it.

As an expositor, Euler has no peers. He produced classic texts in algebra, differential and integral calculus, and the calculus of variations—works that continue to shape the nature of their subjects down to the present day. Further, his writing was fresh and enthusiastic, in contrast to the modern tendency of obscuring a scholar's passion behind the façade of detached, technical prose. Euler was clearly having fun, pursuing the game for its own enjoyment, and exhibiting a pervasive confidence that his quest would be successful.

In beholding such productivity, one is apt to be humble. In all honesty, one is apt to be *overwhelmed*. No author can do justice to the tens of thousands of pages Euler penned over six decades of his career, and it is hard not to feel both inadequate and foolhardy even to consider an undertaking such as this.

Yet his achievements deserve a look. For all the mathematicians who revere Euler's name, relatively few have picked up a volume of the *Opera Omnia* and plunged in. On the contrary, it is the custom of modern mathematicians to learn the subject from textbooks rather than from original sources. Because of changes in notation and emphasis that occur over time, not to mention real advances that can render a prior discussion obsolete, this is not an inherently bad idea.

But something is lost if we deal only in substitutes, only in proxies. Original mathematics, even if centuries old, can be as stirring as the theorems proved last week. This is especially true of Euler's work, as Raymond Ayoub so cogently observed when he wrote:

> Reading his papers is an exhilarating experience; one is struck by the great imagination and originality. Sometimes a result familiar to the reader will take on an original and illuminating aspect, and one wishes that later writers had not tampered with it.[2]

No student of literature would be satisfied with a mere synopsis of *Hamlet*. In like fashion, no mathematician should go through a career without meeting Euler face to face. To do otherwise suggests not only an indifference about the past but also, in some fundamental way, a genuine selfishness.

My ground rules for this book are simple: I focus each chapter upon a subject to which Euler made a significant contribution. Chapters begin with a discussion of what was known prior to Euler; this provides an opportunity to

---

[2]Raymond Ayoub, "Euler and the Zeta Function," *The American Mathematical Monthly*, Vol. 81, No. 10, 1974, p. 1069.

introduce such predecessors as Euclid, Heron, Briggs, and Bernoulli—giants upon whose shoulders Euler would stand. I next examine an Eulerian "great theorem" that pushed the frontiers as only he could. In so doing, I pledge to be as faithful as possible in explaining his original line of attack. Each chapter concludes with an epilogue, either discussing Euler's subsequent work on the topic or describing how later mathematicians further developed his ideas.

As a consequence, this book meanders through number theory, analysis, complex variables, algebra, geometry, and combinatorics—these being but a few of the areas in which Euler made an impact. Selections of theorems— indeed, selections of the areas themselves—are my own. Moreover, because Euler was a master at devising multiple proofs of the same result, one must choose among equally intriguing routes to the same end. Fifty different authors operating under the same ground rules would come up with fifty different books (and I'd be interested to read the other forty-nine). But this one is mine.

What mathematical prerequisites are necessary for the chapters ahead? On the one hand, this volume is not aimed at the rank beginner. Readers should be familiar with such concepts as "integration by parts" or "prime numbers" or "geometric series." I imagine that a few college math courses would provide more than sufficient background for everything I cover.

On the other hand, the book certainly does not assume a graduate school mastery of any branch of mathematics. In a very real sense, that would defeat my purpose. I hope I have made the material accessible to the widest possible audience of "mathematically literate" readers so that it is, in the best sense of the term, *expository*.

As I begin, I make both an observation and a request.

The observation is that Euler was far from infallible. He operated in an era whose standards of mathematical rigor were far more primitive than those of today. As we shall see, some of his arguments were questionable, and others were simply wrong. After all, it was Euler who, without hesitation, introduced expressions like

$$1 + \frac{1}{3} + \frac{1}{5} + \frac{1}{7} + \cdots = 0.66215 + \frac{1}{2}\ln(\infty)\,^3$$

or

$$\frac{1 - x^0}{0} = -\ln x.\,^4$$

---

[3]Leonhard Euler, *Elements of Algebra*, trans. John Hewlett, Springer-Verlag, New York (1840 Reprint), p. 296.
[4]Euler, *Opera Omnia*, Ser. 1, Vol. 14, p. 12.

The modern reader may dismiss these with a knowing smirk, but one dare not laugh too quickly. Because both the left and right sides of the first equation are infinite, it is not really incorrect (even if the 0.66215 on the right seems absurdly superfluous). And the second equation, if slightly modified to read "$\lim_{t \to 0^+}(1 - x^t)/t = -\ln x$ for $x > 0$," makes perfect sense. Here, as often happens with Euler's "mistakes," we come to realize that, though this be mathematical madness, yet there is method in it.

I also make my request, namely that irate readers not complain about the omission of their favorite Eulerian theorem. At the outset, I plead guilty to such charges, for I have omitted virtually *all* of Euler's work. This book represents just the tip of the mathematical iceberg or, perhaps more appropriately, of the mathematical glacier.

At best, I hope to share my personal enthusiasm for a tiny fragment of Euler's remarkable vision. In spite of the passage of centuries, his contributions remain of the highest order, and his impact upon mathematics is everywhere evident. No matter their speciality, mathematicians of today may truly say of Euler what was once said of Wren:

"If you seek his monument, look around."

Biographical Sketch

Euler's life fits snugly within the eighteenth century. Born in the spring of 1707, he lived 76 years until the autumn of 1783. This makes him the close contemporary of another quintessential citizen of that century, Benjamin Franklin (1706–1790). Although of different temperaments, different interests, and even different hemispheres, both Franklin and Euler were widely esteemed in their own day, and both had a profound impact upon the course of Western civilization.

Although focusing primarily on Euler's mathematics, this book should also provide at least a quick survey of his life. In one sense, that life was not especially exciting. Euler was a fairly conventional person, by all accounts kind and generous, but one who lacked the flair of some of his century's better known figures. Unlike Washington (1732–1799), he did not command armies to victory; unlike Robespierre (1758–1794), he did not lead—or succumb to—a political revolution; unlike Captain Cook (1728–1779), he did not sail the seas to explore unknown continents.

Yet in another sense, Euler *was* a great adventurer. His adventures, of course, were of the intellectual sort, carrying him not across the physical world but through a wonderful mathematical landscape. Exploration, after all, can take many forms.

Leonhard Euler was born near Basel, Switzerland. His father was a Protestant clergyman of modest means who entertained the hope that Leonhard would follow him into the pulpit. His mother also came from a pastoral family, so the deck seemed stacked: young Euler appeared destined for the ministry.

He was a precocious youth, blessed with a gift for languages and an extraordinary memory. Euler eventually carried in his head an assortment of curious information, including orations, poems, and lists of prime powers. He also was a fabulous mental calculator, able to perform intricate arithmetical computations without benefit of pencil and paper. These uncommon talents would serve him well later in life.

After entering the University of Basel at age 14, Leonhard encountered its most famous professor, Johann Bernoulli (1667–1748). Two facts about Bernoulli should be noted. First, he was a proud and arrogant man, as quick to demean the work of others as to praise that of himself. Second, any such praise was probably deserved. In 1721, Johann Bernoulli could claim to be the world's greatest active mathematician (Leibniz had died a few years before, and the aged Newton had long since abandoned mathematics). It was only by chance that he found himself in Basel, a small city that was hardly the intellectual capital of the world. Yet there he was when Euler needed a mentor.

Not Euler's teacher in a modern sense of the term, Bernoulli instead became a guide for the young scholar, suggesting mathematical readings and making himself available to discuss those points that seemed especially difficult. As Euler recalled years later,

> I was given permission to visit [Johann Bernoulli] freely every Saturday afternoon and he kindly explained to me everything I could not understand.[1]

Euler asserted that this loose tutorial arrangement was "undoubtedly...the best method to succeed in mathematical subjects," and crusty Johann Bernoulli came to realize that his young tutee was something special. As the years passed and their relationship matured, it was Bernoulli who more and more seemed to become the pupil. Johann, a man not easily given to compliments, once wrote to Euler these generous lines:

> I present higher analysis as it was in its childhood, but you are bringing it to man's estate.[2]

At university, Euler's education was not limited to mathematics. He spoke on the subject of temperance, wrote on the history of law, and eventually completed a master's degree in philosophy. Then, fulfilling his apparent destiny, Euler entered divinity school to study for the ministry.

But the call of mathematics was too strong. He later remembered:

> I had to register in the faculty of theology, and I was to apply myself ... to the Greek and Hebrew languages, but not much progress was

---

[1] Charles C. Gillispie, ed., *Dictionary of Scientific Biography*, Leonhard Euler, p. 468.
[2] Morris Kline, *Mathematical Thought from Ancient to Modern Times*, Oxford U. Press, New York, 1972, p. 592.

Euler on Swiss currency

made, for I turned most of my time to mathematical studies, and by my happy fortune the Saturday visits to Johann Bernoulli continued.[3]

He left the ministry to others—Euler would become a mathematician.

His progress was rapid. At age 20, he earned recognition in an international scientific competition for his analysis of the placement of masts on a sailing ship. This was remarkable for one so young and so landlocked (after all, Euler had spent his entire life in Switzerland). It was the harbinger of successes to come.

Then, as now, it did not hurt to have friends in high places. In 1725, Johann's son Daniel Bernoulli (1700–1782) arrived in Russia to assume a position in mathematics at the new St. Petersburg Academy, and the next year Euler was invited to join him. The only opening at the time was in physiology/medicine, but jobs were scarce, so Euler accepted the offer. Knowing nothing of the medical arts, he set about learning the subject in characteristically industrious fashion—albeit from a somewhat *geometrical* point of view.

Upon his 1727 arrival in St. Petersburg, Euler learned that he had been reassigned to physics rather than physiology—surely a fortuitous development not only for him but also for those patients whom he might have operated upon with compass and straightedge. During the early years in Russia, Euler resided at the home of Daniel Bernoulli, and the two engaged in extended discussions of physics and mathematics that to some extent previewed the course of European science over the coming decades.

---

[3]Clifford Truesdell, "Leonhard Euler, Supreme Geometer," in Euler's *Elements of Algebra*, p. xii.

In 1733, Daniel Bernoulli left for an academic post in Switzerland. On the one hand, the departure of his good friend left a hole in Euler's life. On the other, it opened up the chair in mathematics, which Euler soon occupied.

With such professional advancement, Euler found himself comfortably situated, and soon thereafter he married. His wife was Katharina Gsell (?–1773), daughter of a Swiss painter living in Russia. Over four decades of their happy and productive marriage, the Eulers had 13 children. Unfortunately, as was common at the time, only five survived to adolescence, and only three outlived their parents.

Intellectual life at the St. Petersburg Academy suited Euler perfectly. He devoted a vast amount of effort to research but was constantly at the disposal of the state—which, after all, paid his salary. Time and again he found himself as a scientific consultant to the government, in which capacity he prepared maps, advised the Russian navy, and even tested designs for fire engines. However, he drew the line when asked to cast a horoscope for the young Czar, a job he quickly passed to another.

Meanwhile his fame was growing. One of his earliest triumphs was a solution of the so-called "Basel Problem" that had perplexed mathematicians for the better part of the previous century. The issue was to determine the *exact* value of the infinite series

$$1 + \frac{1}{4} + \frac{1}{9} + \frac{1}{16} + \frac{1}{25} + \cdots + \frac{1}{k^2} + \cdots .$$

Numerical approximations had revealed that the series sums to a number somewhere in the vicinity of 8/5, but the exact answer eluded a string of mathematicians ranging from Pietro Mengoli (1625–1686), who posed the problem in 1644, through Jakob Bernoulli (1654–1705)—Johann's brother and Daniel's uncle—who brought it to the attention of the broader mathematical community in 1689. Well into the next century the problem remained unsolved, and anyone capable of summing the series was certain to make a major splash.

When it happened in 1735, the splash was Euler's.[4] The answer was not only a mathematical *tour de force* but a genuine surprise, for the series sums to $\pi^2/6$. This highly non-intuitive result made the solution all the more spectacular and its solver all the more famous. (Euler's reasoning is described in Chapter 3 of this book).

---

[4]Ron Calinger, "Leonhard Euler: The First St. Petersburg Years (1727–1741)," *Historia Mathematica*, Vol. 23, 1996, pp. 121–166 contains an account of the Basel problem and an excellent survey of Euler's first stay in Russia.

With the Basel problem behind him and the promise of good things ahead, Euler pursued his research at a breathtaking pace. Paper after paper flowed from his pen into the journal of the St. Petersburg Academy, so that for some issues half the articles in the publication were his. He seemed to be living in a mathematician's paradise.

But three problems darkened this period. First was the political turmoil that swirled across Russia in the aftermath of the unexpected death of Catherine I. Her absence left a leadership vacuum that, in conjunction with the suspicions and intrigues of the day, had dangerous consequences. Among these were an intolerance of dissent and a growing suspicion of foreigners. The fact that the Academy was staffed almost exclusively by non-Russians led Euler to describe his situation as "rather awkward."[5]

Second, the Academy was run by a pompous bureaucrat named Johann Schumacher. In the words of Clifford Truesdell, Schumacher's primary occupation lay "in the suppression of talent wherever it might rear its inconvenient head."[6] Although Euler was diplomatic in dealing with his boss, he surely could not have been comfortable working under a martinet with such undeserved self-importance.

The final problem was physical: the deterioration of Euler's eyesight. As early as 1738 he experienced a loss of vision in his right eye. Euler attributed this to overwork, particularly to his intense efforts at cartography, but modern medical opinion suggests it more likely was the result of a severe infection he had recently suffered.

The impact of his visual decline was—in terms of Euler's mathematics—nil. Visual impairment or no, Euler continued his program of research. He wrote about ship-building, acoustics, and the theory of musical harmony. With the encouragement of his friend Christian Goldbach (1690–1764), Euler made seminal discoveries in classical number theory (see Chapter 1) and pushed into the uncharted waters of analytic number theory (see Chapter 4). In response to a letter from Philippe Naudé (1684–1745), he lay the groundwork for the theory of partitions (Chapter 8). And it was during this period that he wrote his text, *Mechanica*, which presented the Newtonian laws of motion within a framework of calculus. For this, the *Mechanica* has been called "a landmark in the history of physics."[7]

---

[5] Truesdell, p. xx.
[6] Ibid., p. xv.
[7] Calinger, p. 143.

With such an output came a matching reputation, which in turn generated an offer from Prussia's Frederick the Great (1712–1786) to become a member of the newly revitalized Berlin Academy. Because of the uneasy political situation in Russia, which Euler described as "a country where every person who speaks is hanged," the offer looked appealing.[8] Thus in 1741 Leonhard, Katharina, and family made the move to Germany.

Berlin was home for a quarter of a century, the middle phase of Euler's mathematical career. During this time he published two of his greatest works— a 1748 text on functions, the *Introductio in analysin infinitorum* (discussed in Chapter 2), and a 1755 volume on differential calculus, the *Institutiones calculi differentialis*. This period also saw him investigate complex numbers and discover "Euler's identity"—$e^{i\theta} = \cos\theta + i\sin\theta$ (see Chapter 5)—as well as offer a proof of the fundamental theorem of algebra (which we treat in Chapter 6).

While in Berlin, Euler was asked to provide instruction in elementary science to the Princess of Anhalt Dessau. The result was a multi-volume masterpiece of exposition, subsequently published as *Letters of Euler on Different Subjects in Natural Philosophy Addressed to a German Princess*.[9] This compilation of over 200 "letters" introduced subjects as diverse as light, sound, gravity, logic, language, magnetism, and astronomy. In the course of the work, Euler explained why it is cold atop a high mountain in the tropics, why the moon looks larger when it rises, and why the sky is blue. He ranged further afield when he discussed the origin of evil, the conversion of sinners, and the intriguing topic of "Electrization of Men and Animals."

Writing about vision in a "letter" dated August 1760, Euler began with these words: "I am now enabled to explain the phenomena of vision, which is undoubtedly one of the greatest operations of nature that the human mind can contemplate."[10] The poignancy of this remark, coming as it did from a partially—and soon to be totally—blind author, is striking. But Euler was not one to let personal misfortune interfere with his attitude toward the wonders of Nature.

Letters to a German Princess became an international hit. The work was translated into a host of languages across Europe and eventually published (in

---

[8] Euler, *Letters of Euler on Different Subjects in Natural Philosophy*, Arno Press, New York, 1975, p. 19.
[9] Item #8 is an English translation of Euler's *Letters to a German Princess*.
[10] Ibid., p. 155.

1833) in the United States. In the preface to the American edition, the publisher gushed over Euler's expository skill in guaranteeing that

> the delight of the reader is, at every step, commensurate with her improvement, and each succeeding acquisition of knowledge becomes a source of still increasing gratification.[11]

In the end, this was Euler's most widely read book. It is not always the case that a scholar working at the very frontier of research can step back to write a treatise accessible to the layman, but this Euler surely did. *Letters to a German Princess* remains to this day one of history's finest examples of popular science.[12]

In spite of the fact that Euler had deserted his colleagues in Russia, they bore him no ill will. From Germany he continued to edit the St. Petersburg journal, to publish article after article in its pages, and to receive a regular stipend from his old employer. Such cordiality continued even through the Seven Years' War, which saw Russian troops invade Berlin. A friendly relationship with St. Petersburg would prove significant in the years to come.

Beyond his mathematical research, Euler was deeply involved in administrative duties at the Berlin Academy. Although not officially the Academy's director, he informally played that role. In the process, he assumed a peculiar array of responsibilities, from juggling budgets to overseeing greenhouses.

But all was not well in Berlin, for Frederick the Great had developed an inexplicable contempt for his most famous scholar-in-residence. The animosity seems to have stemmed as much from a personality conflict as anything. Frederick regarded himself as an erudite, witty savant. He loved philosophy, poetry, and anything French. In fact, affairs at the Academy were conducted in French, not German. To the King, Euler was something of a bumpkin—a brilliant bumpkin to be sure, but a bumpkin all the same. Conventional in his tastes, Euler was a hard-working family man and a devout Protestant. "As long as he preserved his sight," we are told,

> he assembled the whole of his family every evening, and read a chapter of the Bible, which he accompanied with an exhortation. Theology was one of his favourite studies, and the doctrines which he held were the most rigid doctrines of Calvinism.[13]

---

[11] Ibid., p. ii.

[12] See also Ron Calinger, "Euler's letters to a princess of Germany as an expression of his mature scientific outlook," *Archives of the History of the Exact Sciences*, Vol. 15, No. 3, 1975/76, pp. 211–233.

[13] Euler, *Letters of Euler on Different Subjects in Natural Philosophy*, p. 26.

Here was someone of a different breed than the glittering sophisticates at the Berlin Academy. Before long, Frederick took to calling him "my cyclops," a cruel reference to Euler's limited vision.

Making matters worse was the frosty relationship that developed between Euler and the Academy's other superstar, Voltaire (1694–1778). At least for a time, Voltaire enjoyed several advantages in the circle of Frederick the Great—he was celebrated as an author and satirist; he was as sophisticated as the King; and he was thoroughly French. Euler was not spared Voltaire's caustic wit. The latter characterized him as one who "never learnt philosophy" and thus had to satisfy himself "with the fame of being the mathematician who in a given time has filled more sheets of paper with calculations than any other."[14]

Thus, despite bringing to the Berlin Academy a mathematical glory it would never again achieve, Euler was forced out. Matters in Russia had improved during his absence, particularly with the installation of Catherine the Great (1729–1796), so Euler was only too happy to return. The St. Petersburg Academy must have barely believed its good fortune when, in 1766, it welcomed back the greatest mathematician in the world. This time, Euler would stay for good.

Although his scientific life proceeded apace, the next few years brought two personal tragedies. First, he suffered the failure of his remaining good eye. By 1771 Euler was virtually blind. This left him without the ability to write or read anything other than very large characters. Then, late in 1773, Katharina died. Coupled with his recent blindness, this loss could well have marked the end of Euler's productive years.

Euler, however, was no ordinary man. Although unable to see, he not only maintained but even increased his scientific output. In the year 1775, for instance, he wrote an average of one mathematical paper *per week*. Such productivity came in spite of the fact that he now had to have others read him the contents of scientific papers, and he in turn had to dictate his work to diligent scribes. During this descent into blindness, he wrote an influential textbook on algebra, a 775-page treatise on the motion of the moon, and a massive, three-volume development of integral calculus, the *Institutiones calculi integralis*. Never was his remarkable memory more useful than when he could see mathematics only in his mind's eye.

That this blind and aging man forged ahead with such gusto is a remarkable lesson, a tale for the ages. Euler's courage, determination, and utter unwilling-

---

[14]Truesdell, p. xxix.

Portrait of the mature Euler

ness to be beaten serves, in the truest sense of the word, as an inspiration for mathematician and non-mathematician alike. The long history of mathematics provides no finer example of the triumph of the human spirit.

Three years after his wife's death Euler married her half sister, thereby finding a companion with whom to share his last years. These stretched until September 18, 1783. On that day, Euler spent time with his grandchildren and then took up mathematical questions associated with the flight of balloons. This was a topic of interest due to the Montgolfier brothers' recent ascent above Paris in a hot-air balloon—an event witnessed, incidentally, by a diplomat of the new American nation, Benjamin Franklin.[15]

After lunch Euler made some calculations on the orbit of the planet Uranus. Undoubtedly he would have found the behavior of Uranus a rich source of new

---

[15]Roger Burlingame, *Benjamin Franklin: Envoy Extraordinary*, Coward-McCann, New York, 1967, p. 182.

problems. In the decades to come, the planet's peculiar orbit, analyzed in light of equations that Euler had refined, led astronomers to search for—and to discover—the even more distant planet Neptune. Had Euler the time, he would have enjoyed the challenge of seeking a new planet mathematically.

But Euler was not to have such an opportunity. In the late afternoon of that typically busy September day, he was struck down by a massive hemorrhage that caused his immediate death. Mourned by his family, by his colleagues at the Academy, and by the world's scientific community, Leonhard Euler was laid to rest in St. Petersburg. Only then did this great engine of mathematics fall silent.

Euler left behind a legacy of epic proportions. So prolific was he that the journal of the St. Petersburg Academy was still publishing the backlog of his papers a full 48 years after his death. There is hardly a branch of mathematics—or for that matter of physics—in which he did not play a significant role.

In his eulogy, the Marquis de Condorcet observed that whosoever pursues mathematics in the future will be "guided and sustained by the genius of Euler" and asserted, with much justification, that "all mathematicians ... are his disciples."[16]

In the eight chapters that follow, sustained by this genius, we shall examine a tiny fraction of Euler's output. It is only a sampler. But, heeding the advice of Laplace, we shall be reading the work of a master.

---

[16]Euler, *Opera Omnia*, Ser. 3, Vol. 12, p. 308.

CHAPTER 1

Euler and Number Theory

Of all branches of mathematics, none is so natural—nor so deceptively difficult—as the theory of numbers. Its object is to understand the positive integers, surely the most fundamental of mathematical entities. To the uninitiated, number theory seems far simpler than its more sophisticated cousins like trigonometry or calculus. After all, any eight-year old can count to fifty, but how many know the Law of Cosines or the Chain Rule?

It takes very little number theoretic exposure to disabuse the uninitiated of this notion. In fact, the innocent-looking whole numbers are the source of some of the deepest, most vexing problems in mathematics. Hiding their secrets with an embarrassing ease, the integers provide a worthy challenge for the greatest of mathematicians.

Perfect numbers, the subject of this chapter, were of interest as far back as classical times. Euclid (ca. 300 BCE) included a major theorem about such numbers in his masterpiece, the *Elements*, and twenty centuries later Leonhard Euler revisited the topic to finish what Euclid had begun. Yet even Euler left important questions unanswered. To this day, as with so many issues in number theory, the final chapter remains to be written, and the quest for perfect numbers, in the words of Victor Klee and Stan Wagon, "is perhaps the oldest unfinished project of mathematics."[1]

Prologue

Euclid's *Elements* is recognizable even by non-mathematicians as the foremost geometry text of the ancient Greeks. But many are surprised to learn that Euclid devoted three of the thirteen books (or chapters) of the *Elements* to number theory.

---

[1] Victor Klee and Stan Wagon, *Old and New Unsolved Problems in Plane Geometry and Number Theory*, Mathematical Association of America, 1991, p. 178.

This reflects a tradition in Greek thought going back to the Pythagorean philosophers of the sixth century BCE. For them, whole numbers were more than just mathematical abstractions—they were objects of reverence and contemplation, woven into the very fabric of Nature. The Pythagoreans attributed to whole numbers an importance having as much to do with mysticism as with mathematics.

Working within this tradition, Euclid began Book VII of the *Elements* with 22 definitions. Some are easily recognizable today. For instance, Euclid defined a "prime number" to be one that is "measured by a unit alone." Others, like "an even-times odd number"—which Euclid defined as "that which is measured by an even number according to an odd number"—sound quaint to our ears.

The definition of importance for this chapter, and the last on his list, was:

Definition. A **perfect** number is that which is equal to its own parts.

The modern reader may be somewhat confused by the terminology. The matter becomes clearer if we recognize that, for Euclid, "part" meant "proper whole number divisor" and that "equal to" meant "equal to the sum of." With these modifications, we transform Euclid's words into their modern equivalent:

Definition. A whole number is **perfect** if it is equal to the sum of its proper divisors.

For example, the number 6 is perfect because its proper divisors are 1, 2, and 3, and $1 + 2 + 3 = 6$. So too are 28 ($1 + 2 + 4 + 7 + 14 = 28$); 496 ($1 + 2 + 4 + 8 + 16 + 31 + 62 + 124 + 248 = 496$); and 8128 ($1+2+4+8+16+32+64+127+254+508+1016+2032+4064 = 8128$). These four were the only perfect numbers known in ancient Greece, and no others below 10,000 display such "perfection." Clearly they are few and far between.

Nicomachus, a Greek mathematician of the first century, held such numbers in high regard. He observed that perfect numbers are remarkable and rare, "even as fair and excellent things are few ... while ugly and evil ones are widespread."[2] And in later centuries, imaginative scholars attached to perfect numbers a significance of the most outlandish kind. For instance, the number 6 was taken as representing the *perfect* union of the sexes, for $6 = 3 \times 2$, where 3

---

[2]Nicomachus of Gerasa, *Introduction to Arithmetic*, trans. Martin L. D'ooge, U. of Michigan Press, 1938, p. 209.

is a "male" number and 2 is a "female" one (for reasons that should be evident to all but the anatomically challenged). Clearly, our predecessors made perfect numbers carry some pretty heavy baggage.

Euclid bypassed such numerological rubbish and addressed the subject from a purely mathematical viewpoint. Although defining perfect numbers at the outset of Book VII, he never mentioned them again until the end of Book IX—that is, until the final number theoretic proposition in the *Elements*. Undoubtedly, Euclid was saving the best until last, for his theorem was a classic, providing a splendid recipe for perfect numbers.

The result, Proposition 36 of Book IX, was stated by Euclid as:

> If as many numbers as we please beginning from a unit be set out continuously in double proportion, until the sum of all becomes prime, and if the sum multiplied into the last make some number, the product will be perfect.

The modern reader is permitted another blank look. This too needs a bit of translation.

First, the part about beginning with a unit and proceeding in "double proportion" is Euclid's way of describing the series $1 + 2 + 4 + 8 + \cdots$. He supposed that, in continuing this process, the sum turns out to be a prime number; in other words, he assumed that $1 + 2 + 4 + \cdots + 2^{k-1}$ is prime. Then, when this sum is "multiplied into the last"—that is, when we multiply $1 + 2 + 4 + \cdots + 2^{k-1}$ by 2^{k-1} (the "last" term of the progression)—Euclid asserted that the resulting product is a perfect number.

Before examining his proof, we observe that $1 + 2 + 4 + \cdots + 2^{k-1}$ is a finite geometric series which sums to $(2^k - 1)/(2 - 1) = 2^k - 1$. Thus, Euclid's proposition, recast in modern terms, becomes:

Theorem. *If* $2^k - 1$ *is prime and if* $N = 2^{k-1}(2^k - 1)$, *then* N *is perfect.*

Proof. Let $p = 2^k - 1$ be the prime in question. By unique factorization, the proper divisors of $N = 2^{k-1}(2^k - 1) = 2^{k-1}p$ must themselves contain only the primes 2 and p. This means that all such proper divisors can be listed and summed:

Sum of proper divisors of N

$$= 1 + 2 + 4 + \cdots + 2^{k-1} + p + 2p + 4p + \cdots + 2^{k-2}p$$
$$= (1 + 2 + 4 + \cdots + 2^{k-1}) + p(1 + 2 + 4 + \cdots + 2^{k-2})$$

$$= (2^k - 1) + p(2^{k-1} - 1) = p + p2^{k-1} - p$$
$$= p2^{k-1} = N.$$

Because Euclid's number N equals the sum of its proper divisors, it is perfect.

Q.E.D.

Euclid had thereby established a sufficient condition for a number to be perfect. For instance, if $k = 2$, then $2^2 - 1 = 3$ is prime, and so $N = 2(2^2 - 1) = 6$ is perfect. If $k = 3$, then $2^3 - 1 = 7$ is prime, and we get the perfect number $N = 2^2(2^3 - 1) = 28$. And if $k = 13$, we see that $2^{13} - 1 = 8191$ is prime, giving the considerably less obvious example $N = 2^{12}(2^{13} - 1) = 33,550,336$.

This is a fine bit of number theory from 2300 years ago. Not only did Euclid supply a valid proof, but he was able, from the very few perfect numbers known at the time, to discern a pattern. He deserves applause both for mathematical precision and for mathematical *perception*.

Of course, Euclid's theorem replaced one question—finding perfect numbers—with another—finding primes of the form $p = 2^k - 1$. Unfortunately, this new question is anything but easy. Such primes, falling one short of a power of two, have played an important role in number theory. Now called "Mersenne primes" after their seventeenth-century popularizer Marin Mersenne (1588–1648), they are celebrities among primes.

To give a sense of their complexity, we note that if k is composite, then so is $2^k - 1$. This follows from simple algebra, for if $k = ab$,

$$2^k - 1 = (2^a)^b - 1$$
$$= \left[2^a - 1\right]\left[(2^a)^{b-1} + (2^a)^{b-2} + (2^a)^{b-3} + \cdots + (2^a) + 1\right],$$

of which $2^a - 1$ is obviously a factor. For instance, if $k = 6 = 2 \times 3$, we have $2^6 - 1 = (2^2)^3 - 1 = [2^2 - 1][(2^2)^2 + 2^2 + 1]$, which verifies the trivial fact that 63 (i.e., $2^6 - 1$) is divisible by 3 (i.e., $2^2 - 1$) and so is not prime.

This observation allows us to dismiss enormous numbers like $2^{75} - 1$ from among the candidates for Mersenne primes because the exponent is composite. But—and here is where things get complicated—it does not follow that if k is prime, then so is $2^k - 1$. The smallest counterexample is $2^{11} - 1$, a number which, in spite of the prime exponent, factors as $2^{11} - 1 = 2047 = 23 \times 89$.

The quest for Mersenne primes presents a significant challenge. In a 1772 letter to Daniel Bernoulli, Euler claimed to have verified that $2^{31} - 1$ is prime.[3]

---

[3] Leonard Eugene Dickson, *History of the Theory of Numbers*, Vol. 1, G. E. Stechert and Co., New York, 1934, p. 19.

This is the eighth-largest Mersenne prime and, thanks to Euclid's theorem above, generates the perfect number

$$2^{30}(2^{31} - 1) = 2,305,843,008,139,952,128.$$

Early in the nineteenth century, this example was described as

> ... the greatest [perfect number] that will ever be discovered, for, as they are merely curious without being useful, it is not likely that any person will attempt to find one beyond it.[4]

Such pessimism notwithstanding, the search continued. Nowadays, when mathematicians enlist computers to find a new largest prime, invariably they look among numbers of Mersenne's type. Once found, a new megaprime might even get a few inches of space in the daily newspapers, as happened in 1998 when it was announced that $2^{3021377} - 1$ is a (Mersenne) prime.

This discovery, combined with Euclid's ancient result, establishes as a corollary that $2^{3021376}(2^{3021377} - 1)$ is a perfect number—the 37th found as of this writing. The number in question runs to just over 1.8 million digits. To write it out by hand—even at a brisk pace—would consume weeks of (exceedingly dull) work, and then a skeptic might still wish to sum the proper divisors of this behemoth to prove that it really is perfect.

Of course, the skeptic would be wasting his or her time. Euclid's argument long ago settled the issue completely and irrefutably. The skeptic can rest assured that $2^{3021376}(2^{3021377} - 1)$ is a perfect number, for Euclid *proved* it is so. Such is the decisive and eternal power of reason.

Euclid had provided a sufficient condition for a number to be perfect. That is, he proved that *if* a number has a certain form, *then* it will be perfect. It was nowhere claimed that this condition was necessary as well—i.e., *if* a number is perfect, then it must be of the form Euclid described.

Sufficiency and necessity are two very different things. Consider the statement, "If X is an omelette, then X contains eggs." True enough: being an omelette is sufficient to guarantee the object has eggs in it. But egg-containing objects are not *necessarily* omelettes: (consider a quiche, a crepe, or for that matter a chicken). Euclid had provided but half a loaf, which, although better than nothing, fell short of the optimum situation.

The difference between necessity and sufficiency led to an unfortunate error many centuries later. In 1509, Carolus Bovillus (1470–1553) gave a proof

---

[4] Stanley Bezuszka and Margaret Kenney, "Even Perfect Numbers: (Update)^2," *The Mathematics Teacher*, Vol. 90, No. 8, 1997, p. 632.

that *every* perfect number is even.[5] His argument began with a perfect number. Citing Euclid, Bovillus claimed the number must have the form $2^{k-1}(2^k - 1)$, where $(2^k - 1)$ is prime. But such a number has a factor of 2 out front (indeed it has $k - 1$ of them), and so is obviously even.

This "proof" is short, easy, and wrong. By asserting that a perfect number must have the Euclidean structure, Bovillus had confused sufficiency with necessity. His error was the logical equivalent of deducing that a chicken is an omelette.

While on the subject of grievous errors, we note that in 1598 a mathematician named Unicornus (1523–1610) "improved" upon Euclid's theorem by claiming that if k is odd, then $N = 2^{k-1}(2^k - 1)$ is perfect.[6] Among other things, this would guarantee that there are infinitely many perfect numbers, for there certainly are infinitely many odd k. Unfortunately, if $k = 9$, we have $N = 2^8(2^9 - 1) = 130,816$, the sum of whose proper divisors is 171,696. Of course this in no way contradicts Euclid, for $2^9 - 1 = 511 = 7 \times 73$ is not prime. Poor Unicornus had blundered badly, as might be expected of someone named after a mythological creature.

At the dawn of the seventeenth century, Euclid's theorem embodied virtually all that was known of perfect numbers. A complete characterization—necessary and sufficient conditions—remained undiscovered. René Descartes (1596–1650), in a letter to Mersenne of November 15, 1638, stated that every *even* perfect number is "Euclidean"—that is, every even perfect number looks like $2^{k-1}(2^k - 1)$, where $k > 1$ and the expression in parentheses is prime.[7] Unfortunately, we have no record of his reasoning. Whether he devised a proof that was subsequently lost or whether he was just guessing will probably never be known.

This conjecture of Descartes was not only intriguing but also correct. It would remain for another, however, to supply the details.

Enter Euler

For Euler, number theory appears to have been an acquired taste. When a young man, he fell under the spell of differential and integral calculus, then a new and exciting area of research. Mathematicians were enthralled by the

---

[5] Dickson, p. 7.
[6] Ibid., p. 10.
[7] Ibid., p. 12.

power of calculus and its widespread applicability. In modern parlance, the subject was "hot." By comparison, number theory barely registered as a serious mathematical pursuit.

Almost everyone traces Euler's enthusiasm for number theory to a specific proselytizer, Christian Goldbach. He was at the St. Petersburg Academy upon Euler's arrival in 1727 and got to know and appreciate his young colleague. Soon thereafter Goldbach went to Moscow, so he corresponded with Euler by mail. In one such letter, dated December 1, 1729, Goldbach referred to the work of Pierre de Fermat (1601–1665) when he inquired:

Is Fermat's observation known to you, that all numbers $2^{2^n} + 1$ are prime? He said he could not prove it; nor has anyone else done so to my knowledge.[8]

At first, Euler seemed indifferent, but a subsequent, prodding letter from Goldbach sparked his interest. Euler discovered that, on this point, Fermat was wrong, for $2^{2^5} + 1 = 4,294,967,297$ is evenly divisible by 641.[9]

This was just the beginning. For Euler, number theory became a passion. He plunged into Fermat's work, finding it a source of beauty and endless fascination. Over the course of his career, Euler addressed number theoretic matters of profound importance as well as those of considerably less significance. Among the latter was a challenge to find four different whole numbers, the sum of any two of which is a perfect square. With his fearsome foursome of 18530, 38114, 45986, and 65570, Euler supplied a correct, if utterly non-intuitive, answer.[10]

Four volumes of Euler's *Opera Omnia* are devoted to number theory, and many of the results contained therein have become classics. As Harold Edwards has observed, even if this had been Euler's entire mathematical output (and it most surely was *not*), "his contributions to number theory alone would suffice to establish a lasting reputation in the annals of mathematics."[11]

For Euler the matter of perfect numbers arose almost as an afterthought, occupying less than a page of a comprehensive paper "*De numeris amica-bilibus*" in which he considered the so-called amicable numbers.[12] For the

---

[8]Weil, p. 172.

[9]For Euler's argument, see William Dunham, *Journey Through Genius: The Great Theorems of Mathematics*, Wiley, New York, 1990, Chapter 10.

[10]Euler, *Opera Omnia*, Ser. 1, Vol. 5, pp. 330–336.

[11]Harold M. Edwards, *Fermat's Last Theorem*, Springer-Verlag, New York, 1977, p. 39.

record, these are two numbers, m and n, such that the sum of the proper divisors of m is n, and *vice versa*. Amicable pairs are quite rare, the smallest being 220 and 284. In all the centuries prior to Euler, only three pairs had been discovered. He alone, in a veritable explosion of insight, supplied 59 additional pairs!

In the course of his discussion, Euler introduced the following concept, one that would prove useful in the study of amicable pairs and of perfect numbers:

Definition. $\sigma(n)$ is the sum of *all* whole number divisors of n.

(In his paper Euler used the notation $\int n$, but modern authors have replaced the elongated "S" by the lower case Greek "sigma.") Note that where Euclid had summed only the proper divisors of n, Euler found it worthwhile to sum them all. This may seem like an insignificant change, but it opened the door to some crucial observations.

For example, we see that $\sigma(5) = 1+5 = 6$ and $\sigma(6) = 1+2+3+6 = 12$. Clearly, the sum of the *proper* divisors of n is $\sigma(n) - n$. A moment's thought will reveal that, from this perspective, m and n are an amicable pair if and only if they exhibit the beautiful symmetry: $\sigma(m) = m + n = \sigma(n)$.

More germane to the topic at hand are the following characterizations of prime and perfect numbers:

1. p is prime if and only if $\sigma(p) = p + 1$.
2. N is perfect if and only if $\sigma(N) = N + N = 2N$.

We shall need three other important properties:

3. If p is prime, then $\sigma(p^r) = (p^{r+1} - 1)/(p - 1)$.

This follows because the only divisors of a prime power p^r are prime powers p^s with $0 \le s \le r$. Consequently,

$$\sigma(p^r) = 1 + p + p^2 + \cdots + p^r = \frac{p^{r+1} - 1}{p - 1}.$$

In particular, for $N = 2^r$, we have

$$\sigma(N) = \sigma(2^r) = \frac{2^{r+1} - 1}{2 - 1} = 2^{r+1} - 1 = 2(2^r) - 1 = 2N - 1.$$

This shows that a power of 2 is *never* perfect, because for such powers $\sigma(N)$ falls one unit short of the $2N$ required of perfection. Close, but no cigar.

---

[12]Euler, *Opera Omnia*, Ser. 1, Vol. 5, pp. 353–365.

4. If p and q are different primes, then $\sigma(pq) = \sigma(p)\sigma(q)$.

To prove this relationship, note that the only divisors of pq are 1, p, q, and pq itself, and so $\sigma(pq) = 1+p+q+pq = (1+p)+q(1+p) = (1+p)(1+q) = \sigma(p)\sigma(q)$. As a numerical example, note that $\sigma(21) = 1+3+7+21 = 32 = 4 \times 8 = \sigma(3) \times \sigma(7)$.

5. If a and b are relatively prime, then $\sigma(ab) = \sigma(a)\sigma(b)$.

This extension of #4 says that the key requirement is not the primality of a and b but their *relative* primality. So long as a and b have no common factor other than 1, the result of applying σ to their product equals the product of applying σ to them individually. This characteristic—the so-called "multiplicative property"—is central to the argument that follows and indeed to most considerations involving σ. Euler's keen eye spotted it at once.[13]

We shall not supply the proof of #5 (which can be found in any number theory text), but its essence can be distilled from examining the case where $a = p^2$ and $b = qr$, with p, q, and r three different primes (thereby making a and b relatively prime). Here, we can easily list and sum all divisors of ab:

$$\sigma(ab) = \sigma(p^2qr)$$

$$= 1 + p + p^2 + q + pq + p^2q + r + pr + p^2r + qr + pqr + p^2qr$$

$$= (1 + p + p^2) + q(1 + p + p^2) + r(1 + p + p^2) + qr(1 + p + p^2)$$

$$= (1 + p + p^2)(1 + q + r + qr) = (1 + p + p^2)(1 + q)(1 + r)$$

$$= \sigma(p^2)\sigma(q)\sigma(r)$$

$$= \sigma(p^2)\sigma(qr) \qquad \text{by #4}$$

$$= \sigma(a)\sigma(b).$$

In similar fashion, the general theorem is established. Using it, we can quickly determine sums of divisors of any number whose prime factorization we know. For instance, without having to write down all the divisors of 4800, we see that

$$\sigma(4800) = \sigma(2^6 \times 3 \times 5^2) = \sigma(2^6) \times \sigma(3) \times \sigma(5^2) = 127 \times 4 \times 31 = 15,748.$$

Armed with these elementary yet powerful weapons, Euler returned to Euclid's theorem on perfect numbers. He showed that Euclid's sufficiency

---

[13] Euler, *Opera Omnia*, Ser. 1, Vol. 5, pp. 193–195.

condition, when restricted to the *even* perfect numbers, is also necessary. His proof goes as follows:

Theorem. *If N is an even perfect number, then $N = 2^{k-1}(2^k - 1)$, where $2^k - 1$ is prime.*

Proof. Suppose N is even and perfect. Factor out all powers of 2 to write $N = 2^{k-1}b$ where b is odd. Note that $k > 1$ because N is even and thus has at least one 2 in its factorization. Because N is also perfect, we know that

$$\sigma(N) = 2N = 2(2^{k-1}b) = 2^k b.$$

At the same time, because 2^{k-1} and b are relatively prime, #3 and #5 guarantee that

$$\sigma(N) = \sigma(2^{k-1}b) = \sigma(2^{k-1})\sigma(b) = (2^k - 1)\sigma(b).$$

Equating these expressions for $\sigma(N)$ yields $2^k b = (2^k - 1)\sigma(b)$ or simply

$$\frac{2^k}{2^k - 1} = \frac{\sigma(b)}{b}.$$

As Euler observed, the fraction on the left is in lowest terms, for its numerator exceeds its denominator by 1. Whether the fraction on the right is also in lowest terms is not immediately clear. The best Euler could say was that, for some $c \geq 1$,

$$\sigma(b) = c2^k \tag{1.1}$$

and

$$b = c(2^k - 1). \tag{1.2}$$

He then considered two cases involving the value of c.

Case 1. Suppose $c > 1$

By (1.2) each of the whole numbers 1, b, c, and $2^k - 1$ is a divisor of b. We assert something stronger: that they are four *different* divisors of b. To establish this point, we shall show that no pairwise equality can exist among these numbers:

(a) $1 \neq b$, for otherwise $N = 2^{k-1}b = 2^{k-1}$, which is impossible because a power of 2 cannot be perfect (see #3 above).

(b) $1 \neq c$, for in Case 1 we are stipulating that $c > 1$.

(c) $1 \neq 2^k - 1$, for otherwise $2^k = 2$ and so $N = 2^{k-1}b = b$. This would make N an odd number, contradicting the theorem's hypothesis.

(d) $b \neq c$, for if these were equal, then by (1.2), $b = c(2^k - 1) = b(2^k - 1)$ and thus $1 = 2^k - 1$, which returns us to the already-eliminated (c).

(e) $b \neq 2^k - 1$, for otherwise by (1.2), $b = c(2^k - 1) = cb$, which implies $c = 1$, again contradicting the premise of Case 1.

(f) Finally, if $c = 2^k - 1$, then by (1.2) $b = c(2^k - 1) = c^2$, and so b has at least three divisors: 1, c, and c^2, all different because $c > 1$. Consequently, $\sigma(b)$—the sum of *all* divisors of b—must be at least as large as $1 + c + c^2$. On the other hand, by (1.1), $\sigma(b) = c2^k = c[(2^k - 1) + 1] = c[c + 1] = c^2 + c$. We conclude that $c^2 + c = \sigma(b) \geq 1 + c + c^2$, which is absurd. Hence $c \neq 2^k - 1$.

Consolidating the output of (a)–(f), we see that, as claimed, the numbers 1, b, c, and $2^k - 1$ are four different divisors of b. Therefore each appears as a separate summand in calculating $\sigma(b)$. It follows that

$$\sigma(b) \geq 1 + b + c + (2^k - 1) = b + c + 2^k = c(2^k - 1) + c + 2^k \quad \text{by (1.2)}$$

$$= 2^k(c + 1) > c2^k = \sigma(b). \quad \text{by (1.1)}$$

The contradiction $\sigma(b) > \sigma(b)$ seals it: Case 1 is *impossible*. This leaves as the only alternative:

Case 2. $c = 1$

Then by (1.2) we know that $b = c(2^k - 1) = 2^k - 1$ and by (1.1) we have

$$\sigma(b) = c2^k = 2^k = (2^k - 1) + 1 = b + 1.$$

Because $\sigma(b) = b + 1$, we conclude (by #1 above) that b is prime.

In short, we have demonstrated in Case 2 (the only remaining possibility) that if N is an even perfect number, then $N = 2^{k-1}b = 2^{k-1}(2^k - 1)$, where $2^k - 1$ is prime. The necessity of Euclid's condition is thereby established.

Q.E.D.

The argument, although demanding care in chasing down the cases, is elementary. Certainly no extensive knowledge of number theory is required. Euler's insight was to recast the problem in terms of $\sigma(n)$—thereby focusing not upon the sum of *proper* divisors but the sum of all divisors. It seems a simple thing, yet it was decisive. We would do well to remember Truesdell's observation that "Simplicity does not come of itself but must be created."[14] In this sense, Euler was a master of simplification.

With his proof about even perfect numbers, Euler finished the work begun by Euclid so long before. Their joint result—a collaboration spanning two millennia—should rightly be called the "Euclid-Euler Theorem." This name, to be sure, has an alphabetical appeal to it, but it also hyphenates two of the greatest names from the history of mathematics. It is as though Sophocles and Shakespeare had jointly written a play or Phidias and Michelangelo had jointly carved a statue.

Of course no book contains such a play, and no museum holds such a statue. But the Euclid-Euler Theorem *exists*, a timeless monument to its two brilliant creators. In all of mathematics, there is nothing quite like it.

Epilogue

For all that Euclid and Euler discovered about perfect numbers, there are still gaps in our understanding. For instance, no one yet knows whether there are infinitely many of them. According to Euclid's recipe, the infinitude of perfect numbers would follow immediately from the infinitude of Mersenne primes, but the latter problem has itself remained beyond the reach of mathematicians. The abundance of perfect numbers is an open question.

This epilogue will focus on a different, but equally fascinating, mystery. The reader may have noticed that all of the perfect numbers thus far considered (e.g., 6, 28, 496, 8128) are even. So where are the *odd* ones?

At the outset, we calculate $\sigma(n)$ for the first few odd numbers:

$\sigma(3) = 4$	$\sigma(11) = 12$	$\sigma(19) = 20$	$\sigma(27) = 40$
$\sigma(5) = 6$	$\sigma(13) = 14$	$\sigma(21) = 32$	$\sigma(29) = 30$
$\sigma(7) = 8$	$\sigma(15) = 24$	$\sigma(23) = 24$	$\sigma(31) = 32$
$\sigma(9) = 13$	$\sigma(17) = 18$	$\sigma(25) = 31$	$\sigma(33) = 48$

Note that in all cases, $\sigma(N) < 2N$. These odd numbers fall short of perfection.

On an intuitive level this phenomenon makes sense. Unlike an even number, in which one of the proper divisors is already *half* the given number, an odd number will never get such a boost. That is, whereas 496 is divisible by $496/2 = 248$, the biggest proper divisor of 497 is the relatively puny 71.

---

[14] John Fauvel and Jeremy Gray, eds., *The History of Mathematics: A Reader*, Macmillan, London, 1987, p. 461.

To be perfect, all other proper divisors of 496 need only make up the deficit $496 - 248 = 248$—which, of course, they do since $496 = 2^4(2^5 - 1)$. But for 497 to reach perfection, its remaining proper divisors would have to contribute $497 - 71 = 426$, and they don't even come close. Odd numbers seem at a hopeless disadvantage.[15]

After pages of examples, the weary numerical explorer might reasonably conjecture that if N is odd, then $\sigma(N)$ is *always* less than $2N$. This checks out for every odd number up to and including 943, for which $\sigma(943) = 1008 <$ 2×943. Should this phenomenon continue indefinitely, there could be no odd perfect numbers.

Then occurs one of those marvelous surprises with which mathematics is often blessed: $\sigma(945) = 1920 > 2 \times 945$. Here is an odd number the sum of whose proper divisors exceeds the number. This example destroys our conjecture. Moreover, if the sum of the proper divisors of an odd number can fall below the number or fall above it, there is no obvious reason why the sum cannot hit the number squarely upon the head. Odd perfect numbers are back in the running.

Euler himself addressed this matter in his 1747 paper and admitted to being stumped. "Whether ... there are any odd perfect numbers," he observed in a prophetic passage, "is a most difficult (*difficillima*) question."[16]

When *Euler* calls a problem "most difficult," one can be assured that it is. To this day the existence of odd perfect numbers remains unresolved. In spite of heroic efforts by mathematicians and their computers—employing the best of both carbon and silicon—no odd perfect number has ever turned up. Yet no one has proved that such numbers are an impossibility. Mathematician Richard Guy put it well: the existence of odd perfect numbers is "one of the more notorious unsolved problems of number theory."[17]

This is not to say that there has been no progress on the question, for mathematicians have unearthed many properties an odd number must possess if it is to be perfect. As a case in point, consider the following short but clever 1888 proof from J. J. Sylvester (1814–1897).[18]

Theorem. *An odd perfect number must have at least three* different *prime factors.*

---

[15] Along these lines, see Dan Kalman's "A Perfectly Odd Encounter in a Reno Cafe," *Math Horizons*, April, 1996, pp. 5–7.

[16] Euler, *Opera Omnia*, Ser. 1, Vol. 5, p. 355.

[17] Richard Guy, *Unsolved Problems in Number Theory*, Springer-Verlag, New York, 1981, p. 25.

[18] J. J. Sylvester, *Mathematical Papers*, Vol. 4, Chelsea, New York, 1973 (Reprint), pp. 589–590.

Proof. Suppose first that N is an odd perfect number with a single prime factor—in other words, $N = p^r$ where p is an odd prime and $r \geq 1$. Then $2N = \sigma(N)$, and so

$$2p^r = \sigma(p^r) = \frac{p^{r+1} - 1}{p - 1}$$

by #3 above. Consequently $2p^r - p^{r+1} = 1$, a contradiction because the prime p divides evenly into the left-hand side of the equation but not into the right-hand side. Hence an odd perfect number cannot have a single prime factor.

What about exactly two prime factors? Suppose $N = p^k q^r$ is odd and perfect, where $p < q$ are odd primes. By #5 we know that

$$2N = \sigma(N) = \sigma(p^k q^r) = \sigma(p^k)\sigma(q^r).$$

In other words,

$$2N = (1 + p + p^2 + \cdots + p^k) \times (1 + q + q^2 + \cdots + q^r).$$

Divide both sides of this expression by $N = p^k q^r$ and simplify:

$$2 = \left(1 + \frac{1}{p} + \frac{1}{p^2} + \cdots + \frac{1}{p^k}\right) \times \left(1 + \frac{1}{q} + \frac{1}{q^2} + \cdots + \frac{1}{q^r}\right)$$

$$\leq \left(1 + \frac{1}{3} + \frac{1}{9} + \cdots + \frac{1}{3^k}\right) \times \left(1 + \frac{1}{5} + \frac{1}{25} + \cdots + \frac{1}{5^r}\right)$$

because p, being an odd prime, must be at least 3, and q, being a larger odd prime, must be at least 5. Replace these finite geometric series by their (larger) infinite counterparts and sum the latter to get:

$$2 \leq \sum_{i=0}^{\infty} \frac{1}{3^i} \times \sum_{j=0}^{\infty} \frac{1}{5^j} = \frac{3}{2} \times \frac{5}{4} = \frac{15}{8}, \text{ a contradiction.}$$

So, an odd perfect number, if it exists, must possess three or more prime factors.

Q.E.D.

Subsequently, Sylvester proved that an odd perfect number must have at least four, and then at least five different prime factors.[19] The advantage of such theorems is twofold. First, it limits the field of search. A mathematician on the trail of an odd perfect number—in light of the Sylvester's work—need waste

---

[19] Ibid., p. 604 and pp. 611–615.

no time on something like 227,529, whose factorization $3^4 \times 53^2$ contains only two different primes. Automatically this number is disqualified.

More enticingly, theorems like Sylvester's could lead to a proof of non-existence. For, suppose someone proves that odd perfect numbers must satisfy two conditions that are mutually incompatible—e.g., suppose we knew that odd perfect numbers *must* be divisible by 9 but *cannot* be divisible by 3. Then we could conclude with certainty that odd perfect numbers do not exist.

Unfortunately, no one has yet found an incompatibility among the known properties of odd perfect numbers. This is true in spite of the fact that many such properties have been proved, among which are:[20]

1. An odd perfect number cannot be divisible by 105.
2. An odd perfect number must contain at least 8 different prime factors (an extension of Sylvester's work).
3. The smallest odd perfect number must exceed 10^{300}.
4. The second largest prime factor of an odd perfect number exceeds 1000.
5. The sum of the reciprocals of all odd perfect numbers is finite. Symbolically,

$$\sum_{\text{odd perfect}} \frac{1}{n} < \infty.$$

These conditions have a quirky charm, for they identify specific properties of things that may not exist. If we are uncertain whether there is *any* odd perfect number, what are we to make of information about its second largest prime factor? This is like trying to determine the Tooth Fairy's middle name.

Property 5, although as inconclusive as the rest, is of special interest. As we shall see in Chapter 2, the sum of the reciprocals of all whole numbers—the so-called harmonic series—is infinite. So too is the sum of the reciprocals of all the even numbers, or all the odd numbers, or even all the primes (see Chapter 4). Such numbers are plentiful enough that the sum of their reciprocals diverges to infinity.

By contrast, the sum of the reciprocals of all the perfect squares is finite, as we shall see in Chapter 3. Perfect squares are so widely dispersed among the whole numbers that the sum of their reciprocals does not amount to much. In this sense, Property 5 says that odd perfect numbers are more like the squares— i.e., fairly rare. Of course, given that there are *none* below 10^{300}—a number large enough to give a headache to a supercomputer—this is hardly news.

---

[20] Klee and Wagon, pp. 212–213.

Sylvester, surveying the host of properties that must be true of an odd perfect number, believed that the evidence was nearly conclusive. In 1888, he wrote

> ... a prolonged meditation on the subject has satisfied me that the existence of any one such—its escape, so to say, from the complex web of conditions which hem it in on all sides—would be little short of a miracle.[21]

But miracles do happen. In spite of justifiable skepticism, we cannot logically rule out the existence of an odd perfect number. Eric Temple Bell, a number theorist and popular expositor of mathematics, once grumbled, "To say that number theory is mistress of its own domain when it cannot subdue a childish thing like [odd perfect numbers] is undeserved flattery." Indeed the theory of numbers, which Bell called "the last great uncivilized continent of mathematics," is a ready source of humility.[22]

To ward off discouragement, we can always fantasize about the "perfect" ending to this story. Perhaps the existence of odd perfect numbers will finally be resolved by a young genius named Eunice Eubanks. Then we could christen the most alphabetically blessed proposition in all of mathematics: the Euclid-Euler-Eubanks Theorem.

Until that happy day, we must settle for the definitive results of this chapter. Across a gulf of centuries, Euclid and Euler established the exact nature of the *even* perfect numbers. It was, literally, a collaboration for the ages.

---

[21] Sylvester, p. 590.
[22] E. T. Bell, *The Queen of the Sciences*, Williams and Wilkins, Baltimore, 1931, p. 91.

CHAPTER 2

Euler and Logarithms

In 1748 Euler published a two-volume masterpiece, the *Introductio in analysin infinitorum*, destined to become one of the most influential mathematics books of all time. The *Introductio* was Euler's "pre" calculus text—that is, a collection of topics prerequisite to the study of differential and integral calculus. As the title suggests, Euler employed infinite processes in the development of his precalculus, thereby making the book, for today's reader, an odd mixture of the elementary and the sophisticated.

Referring to the *Introductio*, historian Carl Boyer wrote, "It was this work which made the function concept basic in mathematics."[1] Before Euler, analysis was about properties of "curves"; afterwards, it was about properties of "functions." The change was profound, forever altering the mathematical landscape.

Early in Volume 1, Euler introduced his key definition:

> A **function** of a variable quantity is an analytic expression composed in any way whatsoever of the variable quantity and numbers or constant quantities.[2]

This is not the modern concept. In saying a function is an analytic expression, Euler seemed to equate "function" with "formula." He did not distill the essential idea of functionhood: that to each x in the domain there corresponds a unique y in the range. (To be fair, Euler's views on functions broadened later in his career until they approached the modern formulation.)[3] Nonetheless, this analytic definition was a great improvement over the ill-considered, geometric notion of "curve."

---

[1] Carl Boyer, *History of Analytic Geometry*, Scripta Mathematica, New York, 1956, p. 180.

[2] Euler, *Introduction to Analysis of the Infinite*, Book I, trans. John Blanton, Springer-Verlag, New York, 1988, p. 3.

[3] Israel Kleiner, "Evolution of the Function Concept: A Brief Survey," *The College Mathematics Journal*, Vol. 20, No. 4, 1989, pp. 284–289.

Euler's *Introductio* proved extremely significant, affecting later mathematics in content, style, and notation. Boyer recognized this influence when he wrote that it "did for elementary analysis what the *Elements* of Euclid did for geometry"—high praise indeed.[4] E. W. Hobson offered his own accolade by observing:

> Hardly any other work in the history of Mathematical Science gives to the reader so strong an impression of the genius of the author as the *Introductio*.[5]

In the book, Euler did much more than offer an abstract definition of "function." He also spotlighted those functions that have served ever after as the essential building blocks of analysis. He defined the polynomials, the trigonometric functions, and the exponential functions ("simply powers whose exponents are variables"). "From the inverse of these," he continued (in reference to exponentials), "I have arrived at the most natural and fruitful concept of logarithm."[6]

Because of limitations of space, we cannot examine his development of all of these and so shall focus upon one of the most important: the logarithm. This is not an unreasonable choice, for logarithms were among Euler's favorite analytic tools and will appear repeatedly in subsequent chapters of this book.

At the outset, it is important to emphasize that log *tables* had been devised a century before Euler was born. His contribution was a more conceptual one. Euler defined the logarithm *function*, explicitly recognized the inverse nature of logs and exponentials, perceived the significance of what is known as the "natural" logarithm, and applied his results to theoretical matters far removed from mere computation. As was so often the case, Euler inherited a mathematical concept and left his indelible mark upon it.

Prologue

"Logarithms," asserted Pierre-Simon de Laplace (1749–1827), "by shortening the labors, doubled the life of the astronomer."[7] Although the dearth of 140-year-old astronomers suggests that Laplace exaggerated, his observation was apt. By means of logarithms, multiplication and division are reduced to the sim-

---

[4]Boyer, p. 180.
[5]E. W. Hobson, "Squaring the Circle: A History of the Problem," in *Squaring the Circle and Other Monographs*, Chelsea, New York, p. 42.
[6]Euler, *Introduction to Analysis of the Infinite*, Book I, p. vii.
[7]Victor Katz, *A History of Mathematics: An Introduction*, Addison-Wesley, Reading, MA, 1998, p. 420.

pler operations of addition and subtraction, and—as Euler himself stressed—
"logarithms are especially useful in finding intricate roots."[8] A log table was in
its day what the electronic computer is in the modern era: a timesaving device
of unmatched utility.

The term "logarithm" was coined by John Napier (1550–1617) early in
the seventeenth century. Although Napier was the first to grasp the main idea,
it was his associate Henry Briggs (1561–1631) who, over a period of years,
constructed the familiar table of "common" (base-10) logs. Briggs began by
setting $0 = \log 1$ and $1 = \log 10$. (If this seems self-evident, it may come as a
shock to learn that in Napier's original development $0 = \log 10,000,000$.)

So how does one determine, say, $\log 5$? Nowadays, of course, we punch it
into a calculator or, for the technologically backward, look it up in a dusty old
table. But the creators of logarithms had no such options. They had to derive
logs from scratch—a chore that required an understanding of the properties
of logs, a facility with the tedious algorithm for finding square-roots, and
enormous perseverance.

To get a sense of Briggs's method, we shall calculate $\log 5$ in base
10. First note that $\log \sqrt{10} = \log(10^{1/2}) = \frac{1}{2}\log 10 = 0.50000$. Because
$\sqrt{10} \approx 3.1622777$, we have found that $\log 3.1622777 = 0.5000$, at least
approximately.

The same reasoning—and another square root extraction—yields
$0.250000 = \log \sqrt{\sqrt{10}} = \log 1.7782794$. We continue in this fashion,
repeatedly taking square roots and simultaneously halving the logarithms.
Although Briggs carried out his computations by hand to a mind-numbing 30
decimal places, we provide below an abbreviated table of results:

Number	Logarithm
10	1.00000
$3.1622777 = \sqrt{10} = 10^{1/2}$	0.50000
$1.7782794 = \sqrt{\sqrt{10}} = 10^{1/4}$	0.25000
$1.3335214 = \sqrt{\sqrt{\sqrt{10}}} = 10^{1/8}$	0.12500
$\vdots \qquad \vdots$	$\vdots$
$1.0011249 = 10^{1/2048}$	0.00048828
$1.0005623 = 10^{1/4096}$	0.00024414
$1.0002811 = 10^{1/8192}$	0.00012207

---

[8]Euler, *Introduction to Analysis of the Infinite*, Book I, p. 84.

The sieve of numbers in the left-hand column, whose logarithms are known *exactly*, provides a framework of comparison for finding other logs.

We return to log 5. Again, we take repeated square roots:

$$\sqrt{5} = 2.2360680, \sqrt{\sqrt{5}} = 1.4953488,$$

and so on. Eventually the calculations lead us to $5^{1/4096} = 1.0003930$, a number falling between the bottom two values in the left-hand column. (Yes, it's a fine mesh we've gotten ourselves into.)

Our object is to estimate, via linear interpolation, the corresponding logarithm—designated by x—that would appear in the right-hand column, as shown:

$1.0005623 = 10^{1/4096}$	0.00024414
$1.0003930 = 5^{1/4096}$	x
$1.0002811 = 10^{1/8192}$	0.00012207

This leads to the proportion

$$\frac{x - 0.00012207}{0.00024414 - 0.00012207} = \frac{1.0003930 - 1.0002811}{1.0005623 - 1.0002811},$$

from which it follows that $\log(5^{1/4096}) = x = 0.000170646$. Therefore, $\log 5 = 4096(0.000170646) = 0.698966$.

This approximation is pretty good: to six places, $\log 5 = 0.698970$, so we are off by only four parts in a million. Unfortunately, all of this effort on Briggs's part would have yielded only log 5. To determine log 6, or log 5.34, or any of the other entries in a table of logarithms required a repetition of the procedure. This grim reality compels us to accord Henry Briggs equal measures of admiration and pity.

Pity is especially appropriate because, within a generation, mathematicians had discovered a far easier method for the calculation of logarithms. It involved infinite series, a topic at the frontier of research at the time. Among others, Nicholas Mercator (1620–1687), James Gregory (1638–1675), and the incomparable Isaac Newton (1642–1727) found ways to transform complicated expressions into infinite series and use the latter to get excellent approximations of the former.

Newton, for instance, expanded $(1 + x)^r$ as

$$(1 + x)^r = 1 + rx + \frac{r(r - 1)}{2 \cdot 1}x^2 + \frac{r(r - 1)(r - 2)}{3 \cdot 2 \cdot 1}x^3$$
$$+ \frac{r(r - 1)(r - 2)(r - 3)}{4 \cdot 3 \cdot 2 \cdot 1}x^4 + \cdots.$$

According to Newton, this generalized binomial series was valid whether r "is integral or (so to speak) fractional, whether positive or negative."[9] Using it with $x = \frac{1}{5}$ and $r = \frac{1}{2}$, and truncating after just four terms, he could easily approximate a square root such as

$$\sqrt{1.2} = \left(1 + \frac{1}{5}\right)^{1/2}$$

$$\approx 1 + \frac{1}{2}\left(\frac{1}{5}\right) + \frac{\frac{1}{2}\left(\frac{1}{2} - 1\right)}{2 \cdot 1}\left(\frac{1}{5}\right)^2 + \frac{\frac{1}{2}\left(\frac{1}{2} - 1\right)\left(\frac{1}{2} - 2\right)}{3 \cdot 2 \cdot 1}\left(\frac{1}{5}\right)^3$$

$$= 1 + \frac{1}{10} - \frac{1}{200} + \frac{1}{2000} = \frac{2191}{2000} = 1.09550.$$

This is quite close to the five-place value $\sqrt{1.2} \approx 1.09545$.

Using infinite series to approximate square roots is one thing. Using them to find logarithms is quite another. The first steps in this direction were taken by Gregory of St. Vincent (1584–1667) and Alfonso de Sarasa (1618–1667), whose work suggested a link between logarithms and the *area* under portions of a hyperbola. Their discovery—digested, simplified, and viewed within the context of today's calculus—can be explained as follows:

Let $A(x)$ be the area under the hyperbola $y = \frac{1}{t}$ between $t = 1$ and $t = x$ (see Figure 2.1).

Then

$$A(ab) = \int_1^{ab} \frac{1}{t}\, dt = \int_1^a \frac{1}{t}\, dt + \int_a^{ab} \frac{1}{t}\, dt = \int_1^a \frac{1}{t}\, dt + \int_1^b \frac{1}{au}(a\, du),$$

where the second integral has been transformed by the substitution $t = au$. Consequently,

$$A(ab) = \int_1^a \frac{1}{t}\, dt + \int_1^b \frac{1}{u}\, du = A(a) + A(b).$$

In like fashion, the substitution $t = u^r$ yields

$$A(a^r) = \int_1^{a^r} \frac{1}{t}\, dt = \int_1^a \frac{1}{u^r}(ru^{r-1}\, du) = r\int_1^a \frac{1}{u}\, du = rA(a).$$

These properties of the hyperbolic area—namely, $A(ab) = A(a) + A(b)$ and $A(a^r) = rA(a)$—exactly mirror the corresponding properties of logarithms. Clearly something interesting is afoot.

---

[9]Fauvel and Gray, p. 403.

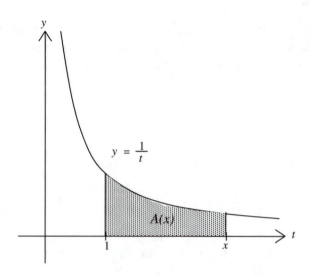

FIGURE 2.1

We now know, of course, that the area in question *is* the so-called natural logarithm, but in the middle of the seventeenth century the connection was not fully understood, and in any case there was no ready means of estimating areas beneath hyperbolas. Resolution of the latter problem, however, was not long in coming. In the 1660s, Mercator and Newton independently approximated these hyperbolic areas—and thus logarithms—by infinite series. Again allowing ourselves the luxury of modern notation, we follow Newton's train of thought.

First, we make the minor modification (essentially a leftward shift of the hyperbola) and define

$$l(1 + x) = \int_0^x \frac{1}{1 + t}\, dt.$$

We then expand $1/(1 + t) = (1 + t)^{-1}$ by Newton's generalized binomial theorem with $r = -1$ and integrate termwise to get

$$l(1 + x) = \int_0^x (1 - t + t^2 - t^3 + t^4 \cdots)\, dt = x - \frac{x^2}{2} + \frac{x^3}{3} - \frac{x^4}{4} + \frac{x^5}{5} \cdots,$$

a simple and beautiful series for hyperbolic area.

Newton recognized that, for small numerical values of x, this series would give accurate approximations of logarithms. In fact, he used it to compute

$$l(1.1) = \int_0^{0.1} \frac{1}{1+t}\, dt$$

to an absurd 57 decimal places and from such approximations showed how to generate a table of base-10 logarithms.[10]

Scholars of the seventeenth century had come a long way. The achievements of Napier and Briggs, of Mercator and Newton, had been impressive indeed. Logarithms were widely used and, with the techniques of infinite series, easily computed. Yet much remained to be done. The unifying theory of the logarithmic function would have to await the next century and its most insightful mathematician.

Enter Euler

We shall consider Euler's development of logarithms from Chapters VI and VII of the *Introductio*. He had previously defined the exponentials—i.e., functions of the form $y = a^z$ where $a > 1$. These he regarded as fairly straightforward. "The extent to which y depends on z," Euler wrote, "is easily understood from the nature of exponents."[11]

He then considered the inverse problem:

> ... we would like to give a value for z such that $a^z = y$. This value of z, insofar as it is viewed as a function of y, is called the **logarithm** of y.

In modern notation, $z = \log_a y$ if and only if $a^z = y$.

Euler had seen the issue in its true light. For him the logarithm was not a mere computational tool but the inverse function of the exponential. And, because there were infinitely many allowable bases, there were infinitely many logarithmic functions to consider.

Later in Chapter VI, he computed $\log_{10} 5 = 0.698970$ using the square-root/sieve method described above. Then, alluding to the series techniques of Newton and Mercator, he observed that "much shorter methods have been found by means of which logarithms can be computed more quickly."[12] These methods he promised to describe in Chapter VII.

---

[10]Derek Whiteside, ed., *The Mathematical Papers of Isaac Newton*, Vol. 2, Cambridge U. Press, 1968, pp. 184–189.

[11]Euler, *Introduction to Analysis of the Infinite*, Book I, p. 77.

[12]Ibid., p. 82.

Euler also stated his "golden rule for logarithms"—namely, that if we have computed $\log_a y$, then "it is an easy task" to find $\log_b y$ where b is any other base.[13] The idea is both simple and powerful. Letting $z = \log_b y$, we have $y = b^z$, and so $\log_a y = \log_a(b^z) = z \log_a b$. Thus $\log_b y = z = \log_a y / \log_a b$.

Not only does this golden rule transform logs from one base to another, but it also implies, as Euler was quick to observe, that the ratio of the logarithms of two numbers is the same no matter what base is used. That is,

$$\frac{\log_b y}{\log_b x} = \frac{\log_a y / \log_a b}{\log_a x / \log_a b} = \frac{\log_a y}{\log_a x}.$$

Euler concluded Chapter VI with a few numerical problems. One, involving interest on debt repayment, could be transferred intact to a modern textbook. Another, on population growth, had a more Biblical—perhaps even antediluvian—flavor:

> Since after the Flood all men descended from a population of six, if we suppose that the population after two hundred years was 1,000,000, we would like to find the annual rate of growth.[14]

It is Chapter VII where Euler's mathematical inventiveness becomes most evident. His objective was to obtain infinite series expansions for the exponential and logarithmic functions, but the elementary nature of his text prohibited use of either differentiation or integration. The resulting argument—in which symbols are flying fast—is as clever as it is non-rigorous.

First he developed a series expansion of the exponential function $y = a^x$, where $a > 1$. To do so, he let ω be an "infinitely small number, or a fraction so small that, although not equal to zero, still $a^\omega = 1 + \psi$, where ψ is also an infinitely small number."[15] To Euler, ω was almost 0, so that $a^\omega \approx a^0 = 1$, the difference being the infinitesimal amount $\psi = a^\omega - 1$.

Euler thus was juggling two infinitely small quantities—ω and ψ. To connect them, he let $\psi = k\omega$, so that $a^\omega = 1 + k\omega$.

At this point, awash in infinite smallness, Euler provided a numerical example to help clarify the situation. He let $a = 10$ and $\omega = 0.000001$, so that $10^{0.000001} = 1 + k(0.000001)$. It follows (from a table of logarithms) that $k = 2.3026$. On the other hand, for $a = 5$ and $\omega = 0.000001$, he found that

---

[13] Ibid., p. 83.
[14] Ibid., p. 86.
[15] Ibid., p. 92.

$k = 1.60944$. "We see," concluded Euler, "that k is a finite number that depends on the value of the base a."[16]

For a finite number x, he now sought the expansion of a^x. Never shy about introducing new variables, he let $j = x/\omega$, so that $a^x = (a^\omega)^{x/\omega} = (1 + k\omega)^j = (1 + kx/j)^j$. This he expanded via Newton's generalized binomial series to get:

$$a^x = 1 + j\left(\frac{kx}{j}\right) + \frac{j(j-1)}{2 \cdot 1}\left(\frac{kx}{j}\right)^2 + \frac{j(j-1)(j-2)}{3 \cdot 2 \cdot 1}\left(\frac{kx}{j}\right)^3$$

$$+ \frac{j(j-1)(j-2)(j-3)}{4 \cdot 3 \cdot 2 \cdot 1}\left(\frac{kx}{j}\right)^4 + \cdots$$

$$= 1 + kx + \frac{j-1}{j}\left(\frac{k^2 x^2}{2 \cdot 1}\right) + \frac{(j-1)(j-2)}{j \cdot j}\left(\frac{k^3 x^3}{3 \cdot 2 \cdot 1}\right)$$

$$+ \frac{(j-1)(j-2)(j-3)}{j \cdot j \cdot j}\left(\frac{k^4 x^4}{4 \cdot 3 \cdot 2 \cdot 1}\right) + \cdots .$$

But x is finite and ω is infinitely small, so $j = x/\omega$ must be infinitely large. It follows—according to Euler—that $(j-1)/j = 1$, $(j-2)/j = 1$, and so on. In modern parlance, he asserted (correctly) that $\lim_{j \to \infty}(j - n)/j = 1$ for any $n \geq 1$, although Euler spoke not of limits but of infinitely small and infinitely large quantities.

In any case, upon eliminating j from the expansion, Euler had arrived at:

$$a^x = 1 + kx + \frac{k^2 x^2}{2 \cdot 1} + \frac{k^3 x^3}{3 \cdot 2 \cdot 1} + \frac{k^4 x^4}{4 \cdot 3 \cdot 2 \cdot 1} + \cdots . \qquad (2.1)$$

He drew two immediate conclusions. First, by letting $x = 1$, he generated a series for the base a in terms of k, namely

$$a = 1 + k + \frac{k^2}{2 \cdot 1} + \frac{k^3}{3 \cdot 2 \cdot 1} + \frac{k^4}{4 \cdot 3 \cdot 2 \cdot 1} + \cdots .$$

Second, because little had thus far been stipulated about a—other than that it be a positive number greater than 1—why not choose a as that *particular* base for which $k = 1$? In other words, initially select the base a so that $a^\omega = 1 + \omega$ when ω is infinitely small. Putting $x = k = 1$ into (2.1) gives an expression for this special base:

$$a = 1 + 1 + \frac{1}{2 \cdot 1} + \frac{1}{3 \cdot 2 \cdot 1} + \frac{1}{4 \cdot 3 \cdot 2 \cdot 1} + \cdots .$$

---

[16]Ibid., p. 93.

This number Euler computed to be *approximately*

$$2.71828182845904523536028,$$

a constant he designated, "[f]or the sake of brevity," by the now-immortal letter
e. The logarithms associated with this base he called "natural or hyperbolic."[17]
Further, with the choice of $k = 1$ and $a = e$, the series in (2.1) became

$$e^x = 1 + x + \frac{x^2}{2 \cdot 1} + \frac{x^3}{3 \cdot 2 \cdot 1} + \frac{x^4}{4 \cdot 3 \cdot 2 \cdot 1} + \cdots = \sum_{r=0}^{\infty} \frac{x^r}{r!},$$

a famous formula indeed.

So far, so good. Next Euler sought a series expansion for the natural
log function (which we shall denote, in modern style, by "ln"). Because, for
infinitely small ω, he knew that $e^\omega = 1 + \omega$, it followed that $\omega = \ln(1 + \omega)$
and that $j\omega = j \ln(1 + \omega) = \ln(1 + \omega)^j$. But ω, although infinitely small, is
nonetheless positive, so "[I]t is clear that the larger the number chosen for j, the
more $(1 + \omega)^j$ will exceed 1."[18] He concluded that, for any positive x, we can
find j so that $x = (1 + \omega)^j - 1$. From this follow three important consequences:

First, $\omega = (1 + x)^{1/j} - 1$.

Second, $1 + x = (1 + \omega)^j = e^{\omega j}$, which implies that $\ln(1 + x) = j\omega$.

Finally, because $\ln(1 + x)$ is finite whereas ω is infinitely small, j must be
infinitely large.

As before, Euler generated an infinite series by employing the binomial
expansion, this time with fractional exponents:

$$\ln(1 + x) = j\omega = j\left[(1 + x)^{1/j} - 1\right] = j\left[1 + \left(\frac{1}{j}\right)x + \frac{\left(\frac{1}{j}\right)\left(\frac{1}{j} - 1\right)}{2 \cdot 1}x^2\right.$$

$$\left. + \frac{\left(\frac{1}{j}\right)\left(\frac{1}{j} - 1\right)\left(\frac{1}{j} - 2\right)}{3 \cdot 2 \cdot 1}x^3 + \cdots\right] - j$$

$$= x - \frac{j - 1}{2j}x^2 + \frac{(j - 1)(2j - 1)}{2j \cdot 3j}x^3 - \frac{(j - 1)(2j - 1)(3j - 1)}{2j \cdot 3j \cdot 4j}x^4 + \cdots.$$

$$(2.2)$$

----

[17] Ibid., p. 97.
[18] Ibid., p. 94.

Again, the infinite magnitude of j guarantees that $(j - 1)/2j = \frac{1}{2}$, that $(2j - 1)/3j = \frac{2}{3}$, that $(3j - 1)/4j = \frac{3}{4}$, and so on. Substituting into (2.2), Euler arrived at the series of Newton and Mercator:

$$\ln(1 + x) = x - \frac{x^2}{2} + \frac{x^3}{3} - \frac{x^4}{4} + \cdots . \tag{2.3}$$

Having come this far, Euler now showed how to *use* the expression of (2.3) to generate tables of logarithms. As it stands, the expansion is of limited value. For instance, if $x = 5$, the series gives $\ln 6 = 5 - 5^2/2 + 5^3/3 - 5^4/4 + \cdots$. "It is difficult to see how can this be," Euler mused, "since the terms of this series continually grow larger and the sum of several terms does not seem to approach any limit."[19]

Not to worry: Euler described a way around this impediment. In the logarithmic series of (2.3), replace x by $-x$ to get

$$\ln(1 - x) = -x - \frac{x^2}{2} - \frac{x^3}{3} - \frac{x^4}{4} - \cdots \tag{2.4}$$

and then subtract series (2.4) from (2.3):

$$\ln(1 + x) - \ln(1 - x) = \left[x - \frac{x^2}{2} + \frac{x^3}{3} - \frac{x^4}{4} + \cdots \right]$$
$$- \left[-x - \frac{x^2}{2} - \frac{x^3}{3} - \frac{x^4}{4} - \cdots \right]$$
$$= 2x + \frac{2x^3}{3} + \frac{2x^5}{5} + \cdots .$$

In other words,

$$\ln \frac{1 + x}{1 - x} = 2 \left[x + \frac{x^3}{3} + \frac{x^5}{5} + \cdots \right]. \tag{2.5}$$

Euler called this series "strongly convergent" for small values of x and observed that it can make the calculation of logarithms astonishingly simple. For instance, we return to $\log_{10} 5$, which we computed above using a Briggsian blizzard of square roots. If $x = \frac{1}{3}$ is substituted into (2.5), we have

$$\ln \frac{1 + \frac{1}{3}}{1 - \frac{1}{3}} = 2 \left[\frac{1}{3} + \frac{1}{81} + \frac{1}{1215} + \frac{1}{15309} + \cdots \right], \text{ or } \ln 2 = 0.693135.$$

Similarly, for $x = \frac{1}{9}$, equation (2.5) yields

---

[19] Ibid., p. 96.

$$\ln\left(\frac{5}{4}\right) = \ln\frac{1 + \frac{1}{9}}{1 - \frac{1}{9}} = 2\left[\frac{1}{9} + \frac{1}{2187} + \frac{1}{295245} + \cdots\right] = 0.223143.$$

Therefore, $\ln 5 = \ln(\frac{5}{4} \times 4) = \ln(\frac{5}{4}) + 2\ln 2 = 0.223143 + 2(0.693135) = 1.609413$, and $\ln 10 = \ln 5 + \ln 2 = 1.609413 + 0.693135 = 2.302548$. Invoking Euler's "golden rule" of logarithms, we conclude that

$$\log_{10} 5 = \frac{\ln 5}{\ln 10} = \frac{1.609413}{2.302548} = 0.698970.$$

To six places, this is exactly the value obtained by Briggs's sieve. But whereas the earlier approach required over two dozen square root extractions, the series attack on the same problem has nary a square root in sight! The method of infinite series is clearly superior, a sign of unquestioned mathematical progress. One is reminded of James Gregory's comment that the power of all previous methods has the same ratio to that of infinite series as the glimmer of dawn has to the splendor of the noonday sun.[20]

However, Euler had loftier aims for the log series than merely computing tables. For one thing, the expansion above was critical in his derivation of an important formula from differential calculus.

First, a word of comparison: the modern development of the log series applies calculus to generate the expansion in (2.3); by contrast, we have seen Euler arrive at this series without explicit use of differential or integral calculus in his derivation (as was necessary due to the "elementary" nature of the *Introductio*). Thus, without the risk of circular reasoning, he was now free to apply the series to problems of calculus.

This he did in his 1755 textbook, the *Institutiones calculi differentialis*. The problem was to find the differential of $\ln x$. His argument, with a slight adjustment in notation, proceeded as follows:[21]

If $y = \ln x$, then its differential is $dy = \ln(x + dx) - \ln x$. (Note: dy is the numerator $f(x + h) - f(x)$ in today's differential quotient, with our h playing the part of Euler's dx.) Exploiting the rules of logarithms and (2.3), Euler wrote:

$$dy = \ln(x + dx) - \ln x = \ln\left(\frac{x + dx}{x}\right) = \ln\left(1 + \frac{dx}{x}\right)$$

$$= \left(\frac{dx}{x}\right) - \frac{(dx/x)^2}{2} + \frac{(dx/x)^3}{3} - \frac{(dx/x)^4}{4} + \cdots.$$

---

[20] Joseph Hofmann, *Leibniz in Paris: 1672–1676*, Cambridge U. Press, 1974, p. 215.

[21] Euler, *Opera Omnia*, Ser. 1, Vol. 10, p. 122.

To Euler it was evident that squares, cubes, and higher powers of the infinitesimal dx are insignificant compared to the infinitesimal itself. Thus, "because all terms after the first vanish," he concluded $dy = dx/x$, which is immediately transformed into the differentiation formula $D_x[\ln x] = dy/dx = 1/x$. Nothing to it.

Such reasoning hails from a pre-rigorous era. This is not to say, however, that it should be casually dismissed. On the contrary, it accurately reflects the standards of its day and, in that context, is both clever and compelling. Given its provenance, this is a derivation worthy of attention.

For Euler, logarithms were one of the chief tools of analysis. Time and again they appeared in his work, often in the most unexpected places (as we shall see in later chapters of this book). The logarithm function, which Euler called a "most natural and fruitful concept," was here to stay.

Epilogue

To conclude the chapter, we shall describe how Euler recognized a link between logarithms and the harmonic series. In the process, he discovered one of the most pervasive, and most perplexing, constants in all of mathematics.

Our story begins with the harmonic series $\sum_{k=1}^{\infty} 1/k$. The apparent simplicity of this series masks its surprising nature. The individual summands 1, $\frac{1}{2}$, $\frac{1}{3}$, ... shrink away to zero, so that, figuratively speaking, the rate at which the sum grows seems to be grinding to a halt. For instance,

$$\sum_{k=1}^{20} \frac{1}{k} \approx 3.60, \quad \sum_{k=1}^{220} \frac{1}{k} \approx 5.98, \quad \text{and} \quad \sum_{k=1}^{20220} \frac{1}{k} \approx 10.49.$$

Note that the sum of the first 20 terms exceeds the sum of the next 200 terms, and this in turn outpaces the contribution of the next 20,000 terms. It is often said that the harmonic series grows with "glacial" slowness. Because of this phenomenon, one could easily believe that the series has some upper bound beyond which it can never pass.

Yet the harmonic series diverges to infinity. That is, its sum grows larger than any preassigned quantity, even though its individual terms tend to zero. This property, which often seems to defy a beginner's intuition, makes the harmonic series one of the first "pathological counterexamples" in analysis. Summing it to infinity is rather like getting something for nothing—or, to be more precise, it is like getting *everything* for nothing.

This behavior had been recognized long before Euler was born. Among the early divergence proofs, an especially elegant one is due to Jakob Bernoulli, older brother of Euler's mentor. It appeared in his 1689 classic *Tractatus de seriebus infinitis (Treatise on infinite series)*, a beautiful treatment of series as they were understood in the generation before Euler.

Jakob Bernoulli's proof, slightly streamlined, ran as follows:[22]

Theorem. *The harmonic series diverges.*

Proof. First we assert that, if $a > 1$, then

$$\frac{1}{a} + \frac{1}{a+1} + \frac{1}{a+2} + \cdots + \frac{1}{a^2} \geq 1.$$

To establish this, consider the sum

$$\frac{1}{a+1} + \frac{1}{a+2} + \cdots + \frac{1}{a^2}.$$

Because it consists of $a^2 - a$ fractions, each greater than the right-handmost, we know that

$$\frac{1}{a+1} + \frac{1}{a+2} + \cdots + \frac{1}{a^2} \geq \frac{1}{a^2} + \frac{1}{a^2} + \cdots + \frac{1}{a^2} = (a^2 - a)\frac{1}{a^2} = 1 - \frac{1}{a}.$$

Adding $1/a$ to both sides proves the assertion.

From this, Bernoulli deduced that the harmonic series can be decomposed into infinitely many pieces of the form

$$\frac{1}{a} + \frac{1}{a+1} + \frac{1}{a+2} + \cdots + \frac{1}{a^2},$$

each totalling one or more. That is,

$$\sum_{k=1}^{\infty} \frac{1}{k} = 1 + \left(\frac{1}{2} + \frac{1}{3} + \frac{1}{4}\right) + \left(\frac{1}{5} + \frac{1}{6} + \cdots + \frac{1}{25}\right)$$

$$+ \left(\frac{1}{26} + \frac{1}{27} + \cdots + \frac{1}{676}\right) + \cdots$$

$$\geq 1 + 1 + 1 + 1 + \cdots$$

It follows the harmonic series grows larger than any finite quantity.   Q.E.D.

---

[22]Jakob Bernoulli, *Ars Conjectandi*, Impression Anastaltique Culture et Civilisation, Brussels, 1968 (Reprint), p. 251.

Something so bizarre was a source of fascination in the seventeenth century. For instance, Gottfried Wilhelm Leibniz (1646–1716) came to believe that English mathematicians had discovered a simple formula for the *partial* sum of the harmonic series: $\sum_{k=1}^{n} 1/k$. Such a formula, which Leibniz had been unable to find, would have been the counterpart of the summation formula for a finite *geometric* series—namely

$$\sum_{k=0}^{n} a^k = \frac{a^{n+1} - 1}{a - 1}$$

—that we exploited in the previous chapter. So eager was Leibniz to acquire information about the British discovery that he offered to provide details of his own derivation of

$$\frac{\pi}{4} = 1 - \frac{1}{3} + \frac{1}{5} - \frac{1}{7} + \cdots$$

in exchange for the harmonic series result.[23] This was not unlike trading mathematical baseball cards. No exchange was forthcoming, of course, because the English mathematicians had no such formula.

Not surprisingly, Euler too was drawn to the harmonic series. In the *Introductio*, he gave his own divergence proof, although one considerably less satisfying than that of Jakob Bernoulli:

Theorem. *The harmonic series diverges.*

Proof. Euler based his brief argument upon the expansion of (2.4) above.[24] That is, letting $x = 1$ in the series

$$\ln(1 - x) = -x - \frac{x^2}{2} - \frac{x^3}{3} - \frac{x^4}{4} - \cdots,$$

he concluded that

$$\ln 0 = -\left(1 + \frac{1}{2} + \frac{1}{3} + \frac{1}{4} + \frac{1}{5} + \frac{1}{6} + \frac{1}{7} \cdots\right)$$

and so

$$1 + \frac{1}{2} + \frac{1}{3} + \frac{1}{4} + \frac{1}{5} + \frac{1}{6} + \frac{1}{7} \cdots = -\ln 0 = \ln(0^{-1}) = \ln\left(\frac{1}{0}\right)$$

$$= \ln \infty = \infty,$$

"because the log of an infinite number is infinite." Q.E.D.

---

[23] Hofmann, p. 33.
[24] Euler, *Introduction to Analysis of the Infinite*, Book I, pp. 234–235.

But Euler noticed something more—a strange and provocative connection between the harmonic series and logarithms. He began by substituting $x = 1/n$ into series (2.3):

$$\ln\left(1 + \frac{1}{n}\right) = \frac{1}{n} - \frac{1}{2n^2} + \frac{1}{3n^3} - \frac{1}{4n^4} - \cdots .$$

Therefore, it was clear that

$$\frac{1}{n} = \ln\left(\frac{n+1}{n}\right) + \frac{1}{2n^2} - \frac{1}{3n^3} + \frac{1}{4n^4} - \cdots , \qquad (2.6)$$

and so for large n the reciprocal $1/n$ is roughly equal to $\ln[(n + 1)/n]$. This suggested to Euler that summing the harmonic series would eventually look a great deal like summing logarithms. He was on his way to an important discovery.

Substituting $n = 1, 2, 3, \ldots$ into (2.6), he got:

$$1 = \ln 2 + \frac{1}{2} - \frac{1}{3} + \frac{1}{4} - \cdots$$

$$\frac{1}{2} = \ln\left(\frac{3}{2}\right) + \frac{1}{8} - \frac{1}{24} + \frac{1}{64} - \cdots$$

$$\frac{1}{3} = \ln\left(\frac{4}{3}\right) + \frac{1}{18} - \frac{1}{81} + \frac{1}{324} - \cdots$$

$$\vdots \quad \vdots \quad \vdots \quad \vdots$$

$$\frac{1}{n} = \ln\left(\frac{n+1}{n}\right) + \frac{1}{2n^2} - \frac{1}{3n^3} + \frac{1}{4n^4} - \cdots .$$

Euler then added down the columns to conclude:

$$\sum_{k=1}^{n} \frac{1}{k} = \left[\ln 2 + \ln\frac{3}{2} + \ln\frac{4}{3} + \cdots + \ln\left(\frac{n+1}{n}\right)\right]$$

$$+ \frac{1}{2}\left[1 + \frac{1}{4} + \frac{1}{9} + \cdots + \frac{1}{n^2}\right] - \frac{1}{3}\left[1 + \frac{1}{8} + \frac{1}{27} + \cdots + \frac{1}{n^3}\right]$$

$$+ \frac{1}{4}\left[1 + \frac{1}{16} + \frac{1}{81} + \cdots + \frac{1}{n^4}\right] - \cdots .$$

The sum of the logs within the first set of square brackets is the log of their product—i.e., $\ln(n + 1)$. Euler approximated the remaining series numerically and in this manner arrived at the estimate[25]

$$\sum_{k=1}^{n} \frac{1}{k} \approx \ln(n + 1) + 0.577218.$$

Consequently, for large n, the partial sum of the harmonic series is like a logarithm plus a constant somewhat bigger than 0.577. Today, this number is denoted by the lower-case Greek letter γ and called—appropriately enough—"Euler's constant." Its precise definition is

$$\gamma = \lim_{n \to \infty} \left[\sum_{k=1}^{n} \frac{1}{k} - \ln(n + 1) \right].$$

Faced with such a concept, a modern mathematician feels compelled to prove that a number so defined actually *exists* (a matter of no small import). For the sake of completeness, we provide a proof:

Theorem. $\lim_{n \to \infty} \left[\sum_{k=1}^{n} \frac{1}{k} - \ln(n + 1) \right]$ *exists.*

Proof. Let

$$c_n = \sum_{k=1}^{n} \frac{1}{k} - \ln(n + 1)$$

and make two observations: First

$$c_{n+1} - c_n = \left[\sum_{k=1}^{n+1} \frac{1}{k} - \ln(n + 2) \right] - \left[\sum_{k=1}^{n} \frac{1}{k} - \ln(n + 1) \right]$$

$$= \frac{1}{n + 1} - \ln(n + 2) + \ln(n + 1)$$

$$= \frac{1}{n + 1} - \int_{n+1}^{n+2} \frac{1}{x} \, dx > 0,$$

because, as seen in Figure 2.2, the integral is the shaded area beneath the hyperbola $y = 1/x$ whereas $1/(n + 1)$ is the larger rectangular area encompassing it. It follows that $c_1 < c_2 < \cdots < c_n < c_{n+1} < \cdots$, so the sequence $\{c_n\}$ is increasing.

---

[25] Euler, *Opera Omnia*, Ser. 1, Vol. 14, pp. 93–95.

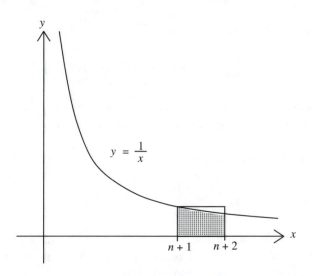

FIGURE 2.2

Second, it is clear from Figure 2.3 that the sum of the rectangular blocks is less than the corresponding area under the curve. Therefore,

$$\sum_{k=1}^{n} \frac{1}{k} = 1 + \sum_{k=2}^{n} \frac{1}{k} < 1 + \int_{1}^{n} \frac{1}{x} \, dx = 1 + \ln n < 1 + \ln(n + 1).$$

Hence

$$c_n = \sum_{k=1}^{n} \frac{1}{k} - \ln(n + 1) < 1 \quad \text{for all } n.$$

Together, these observations establish $\{c_n\}$ as an increasing sequence bounded above by 1. The completeness property of the real numbers guarantees that $\gamma = \lim_{n \to \infty} c_n$ exists.

Q.E.D.

As a quick aside, we note that the definition of Euler's constant found in modern textbooks is the slightly modified

$$\gamma = \lim_{n \to \infty} \left[\sum_{k=1}^{n} \frac{1}{k} - \ln n \right].$$

The change from Euler's original "$\ln(n + 1)$" to today's "$\ln n$" makes no difference whatever because

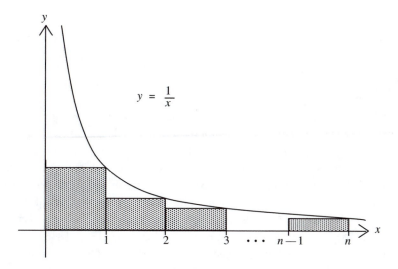

$$y = \frac{1}{x}$$

FIGURE 2.3

$$\lim_{n\to\infty}\left[\sum_{k=1}^{n}\frac{1}{k} - \ln n\right] = \lim_{n\to\infty}\left[\sum_{k=1}^{n}\frac{1}{k} - \ln(n+1) + \ln(n+1) - \ln n\right]$$

$$= \lim_{n\to\infty}\left[\sum_{k=1}^{n}\frac{1}{k} - \ln(n+1)\right] + \lim_{n\to\infty}\ln\left(1 + \frac{1}{n}\right)$$

$$= \gamma + 0 = \gamma.$$

Along with its better-known cousins π and e, the number γ ranks among the most important constants of mathematics, one that Euler endorsed as "worthy of serious attention."[26] Like π and e, it puts in a surprise appearance now and then. It is central to an understanding of the gamma function in higher analysis and figures in such beautiful yet peculiar formulas as

$$\gamma = -\int_{0}^{\infty} e^{-x} \ln x \, dx$$

or

$$\gamma = \left[\frac{1}{2\cdot 2!} - \frac{1}{4\cdot 4!} + \frac{1}{6\cdot 6!} - \cdots\right] - \int_{1}^{\infty}\frac{\cos x}{x}\,dx$$

---

[26]Euler, *Opera Omnia*, Ser. 1, Vol. 15, p. 116.

or

$$\gamma = \lim_{x \to 1^+} \sum_{n=1}^{\infty} \left(\frac{1}{n^x} - \frac{1}{x^n} \right),$$

the last exhibiting a delightful symmetry in x and n.

Like so many profound ideas of mathematics, Euler's constant has been reluctant to give up all of its secrets. For instance, the Italian geometer Lorenzo Mascheroni (1750–1800), in a work titled *Adnotationes ad calculum integrale Euleri*, computed γ to an impressive 32-place accuracy. A few years later, Johann Georg von Soldner (1776–1833) published a value of γ differing from Mascheroni's at the *twentieth* decimal place—a situation that was deemed slightly embarrassing. No less a mathematician than Carl Friedrich Gauss (1777–1855) requested a third party, one F. B. G. Nicolai (1793–1846), whom he described as "an indefatigable calculator," to settle the matter.[27] This Nicolai did, determining the constant to 40 places and thereby showing that von Soldner was right and Mascheroni was wrong.

This mini-crisis over the approximation of a well-defined constant reminds us how far we have come. When computers routinely calculate a few hundred million places of π, a disagreement over the twentieth place of γ seems almost laughable.

It was Mascheroni, by the way, who introduced the symbol γ for this special number. In spite of the fact that he had actually *miscalculated* it, γ is sometimes known as the "Euler-Mascheroni constant." In light of the circumstances, it seems unjust that Mascheroni has been so gloriously hyphenated.

The most enduring mystery about Euler's constant is also one of the most fundamental: is γ rational or irrational? Euler himself said it was a "question of great moment" to determine the character of this number.[28] Yet the basic issue of its rationality/irrationality has thus far stumped the mathematical community. It remains an unresolved problem.

This is in spite of the fact that everyone knows what the answer will be. Something as complicated as γ is not about to be a rational number, a simple fraction with repeating decimal expansion. But, if its irrationality is universally embraced, it has never been proved. Like the existence of odd perfect numbers, the irrationality of γ is a fitting challenge for anyone hoping

---

[27] J. W. L. Glaisher, "On the History of Euler's Constant," *The Messenger of Mathematics*, Vol. 1, 1872, p. 29.

[28] Euler, *Opera Omnia*, Ser. 1, Vol. 15, p. 122.

to achieve mathematical immortality. Potential challengers, however, should be forewarned: this problem has defeated some of the finest minds of the last few centuries, and there are surely easier ways to become famous.

The chapter is at its end. We have seen Euler, who recognized logarithms as functions, develop the series for $\ln(1 + x)$ by generous applications of the infinitely large and infinitely small. This revealed to him a link between logarithms and the harmonic series, which in turn led him to his celebrated constant γ. It is a rambling story, a circuitous chain of ideas. And like so many chains in mathematics, it leads directly through Leonhard Euler.

In the next chapter, we return to the harmonic series, make a small modification as suggested by Jakob Bernoulli, and describe one of Euler's greatest discoveries—his explicit summation of the infinite series

$$\sum_{k=1}^{\infty} \frac{1}{k^2}.$$

CHAPTER 3

Euler and Infinite Series

When the seventeenth century dawned, infinite series were little understood and infrequently encountered. By the century's end, a vast body of specific examples and general theorems had been developed. Jakob Bernoulli's *Tractatus de seriebus infinitis* of 1689, mentioned in the previous chapter, presented a state-of-the-art account of this explosion of knowledge. It was an exciting time, and mathematicians had reason to be proud of their progress over the past hundred years.

Such achievements notwithstanding, there were major problems that defied solution and thus served as conspicuous challenges to scholars of the coming century. Euler, of course, was such a scholar, and in one famous case—the so-called "Basel problem"—he rose to the challenge in spectacular fashion. In this chapter we tell the story of his mathematical triumph.

Prologue

Jakob Bernoulli loved infinite series. Not only did he prove the divergence of the harmonic series, but he also knew exact sums for a number of convergent ones. Simplest among these was the the summation formula for the infinite geometric series:

$$a + ar + ar^2 + \cdots + ar^k + \cdots = \frac{a}{1-r}$$

provided $-1 < r < 1$.

Other, more sophisticated examples could be summed as well. For instance, consider $1 + \frac{1}{3} + \frac{1}{6} + \frac{1}{10} + \frac{1}{15} + \cdots$, where the kth denominator is the so-called kth triangular number, $k(k + 1)/2$. A seventeenth-century evaluation

39

of this series was short and sweet:

$$1 + \frac{1}{3} + \frac{1}{6} + \frac{1}{10} + \frac{1}{15} + \cdots$$

$$= 2\left[\frac{1}{2} + \frac{1}{6} + \frac{1}{12} + \frac{1}{20} + \frac{1}{30} + \cdots\right]$$

$$= 2\left[\left(1 - \frac{1}{2}\right) + \left(\frac{1}{2} - \frac{1}{3}\right) + \left(\frac{1}{3} - \frac{1}{4}\right) + \left(\frac{1}{4} - \frac{1}{5}\right) + \cdots\right]$$

$$= 2[1] = 2,$$

because, within the second set of square brackets, all terms but the first cancel one another. Students of calculus should recognize

$$\sum_{k=1}^{\infty} \frac{1}{k(k+1)/2} = 2$$

as a well-known "telescoping series."

Less familiar is Jakob Bernoulli's summation of the infinite series

$$\frac{a}{b} + \frac{a+c}{bd} + \frac{a+2c}{bd^2} + \frac{a+3c}{bd^3} + \cdots,$$

whose numerators form an arithmetic progression

$$a, a+c, a+2c, a+3c, \ldots$$

and whose corresponding denominators form a geometric progression

$$b, bd, bd^2, bd^3, \ldots.$$

For instance, if $a = 1$, $b = 3$, $c = 5$, and $d = 7$, we have

$$\frac{1}{3} + \frac{6}{21} + \frac{11}{147} + \frac{16}{1029} + \frac{21}{7203} + \frac{26}{50421} + \cdots,$$

whose exact sum is far from obvious.

In Section XIV of the *Tractatus*, Jakob evaluated this series.[1] His insight was to decompose it as follows:

$$\frac{a}{b} + \frac{a+c}{bd} + \frac{a+2c}{bd^2} + \frac{a+3c}{bd^3} + \cdots$$

$$= \left(\frac{a}{b} + \frac{a}{bd} + \frac{a}{bd^2} + \frac{a}{bd^3} + \cdots\right)$$

---

[1] Jakob Bernoulli, p. 247.

$$+ \left(\frac{c}{bd} + \frac{c}{bd^2} + \frac{c}{bd^3} + \cdots \right)$$

$$+ \left(\frac{c}{bd^2} + \frac{c}{bd^3} + \cdots \right)$$

$$+ \left(\frac{c}{bd^3} + \cdots \right)$$

$$\vdots \quad \vdots \quad \vdots \qquad \vdots \qquad \vdots \quad \vdots \quad \vdots$$

Each infinite series in parentheses is geometric and, provided $d > 1$, convergent. Replacing these by their sums gives:

$$\frac{a}{b} + \frac{a+c}{bd} + \frac{a+2c}{bd^2} + \frac{a+3c}{bd^3} + \cdots$$

$$= \frac{a/b}{1 - 1/d} + \frac{c/bd}{1 - 1/d} + \frac{c/bd^2}{1 - 1/d} + \cdots$$

$$= \frac{ad}{bd - b} + \frac{c}{b(d-1)} \left[1 + \frac{1}{d} + \frac{1}{d^2} + \frac{1}{d^3} + \cdots \right]$$

$$= \frac{ad}{bd - b} + \frac{c}{b(d-1)} \left[\frac{1}{1 - 1/d} \right] = \frac{ad^2 - ad + cd}{bd^2 - 2bd + b},$$

because the series in square brackets is geometric as well.

So, for the example above, we have:

$$\frac{1}{3} + \frac{6}{21} + \frac{11}{147} + \frac{16}{1029} + \frac{21}{7203} + \frac{26}{50421} + \cdots = \frac{77}{108}.$$

And there were others. For instance, Jakob found that

$$\sum_{k=1}^{\infty} \frac{k^2}{2^k} = 6$$

and

$$\sum_{k=1}^{\infty} \frac{k^3}{2^k} = 26$$

(which remain good problems to this day).[2] With each success, he must have felt ever more confident of his powers.

Eventually he turned his attention to series of the form

$$\sum_{k=1}^{\infty} \frac{1}{k^p} = 1 + \frac{1}{2^p} + \frac{1}{3^p} + \frac{1}{4^p} + \cdots + \frac{1}{k^p} + \cdots,$$

---

[2] Ibid., pp. 248–249.

which are today called "*p*-series" for obvious reasons. If $p = 1$, we have the (divergent) harmonic series which Jakob had handled perfectly. But what if $p = 2$? What is the exact sum of the series

$$1 + \frac{1}{4} + \frac{1}{9} + \frac{1}{16} + \cdots + \frac{1}{k^2} + \cdots ?$$

The problem was not a new one. Decades earlier, Pietro Mengoli had raised the question and found himself unable to determine this sum. The same could be said for Leibniz, inventor of calculus and master of so many infinite series. Now it was Jakob Bernoulli's turn. One imagines his growing frustration with a series that, on the face of it, seemed no more difficult than those whose secrets he had previously uncovered.

This is not to say that progress was nonexistent. By employing the inequality $2k^2 \geq k(k + 1)$, Bernoulli recognized that

$$\frac{1}{k^2} \leq \frac{1}{k(k + 1)/2},$$

and thus

$$1 + \frac{1}{4} + \frac{1}{9} + \frac{1}{16} + \cdots + \frac{1}{k^2} + \cdots \leq 1 + \frac{1}{3} + \frac{1}{6} + \frac{1}{10} + \cdots + \frac{1}{k(k + 1)/2} + \cdots ,$$

where this latter (telescoping) series converges to 2, as seen above. Because the larger series has a finite sum, Bernoulli reasoned that the smaller one must as well. More explicitly, it was clear that $\sum_{k=1}^{\infty} 1/k^2 \leq 2$. And because $1/k^p \leq 1/k^2$ for all $p \geq 2$, the same argument established that $\sum_{k=1}^{\infty} 1/k^p$ converges for $p = 3, 4, 5, \ldots$.

This stands as an early—and nicely done—example of what is now called the "comparison test" for series convergence. For all of its cleverness, however, it did not provide an *exact* sum for the series in question. On this more difficult matter, Jakob admitted defeat. Writing from Basel, he included in the *Tractatus* his plea for help:

> If anyone finds and communicates to us that which thus far has eluded our efforts, great will be our gratitude.[3]

With these words, the mathematical community was handed a formal, and formidable, challenge. In the end, the "Basel problem" would outlive Jakob Bernoulli and the century that spawned it. Only in the eighteenth century did this great problem meet its match.

---

[3] Ibid., p. 254.

Enter Euler

It is not clear exactly when Euler first considered the matter, but by 1731, at the age of 24, he was hard at work upon it. It occurred to him, as it had to others before, that a reasonable first step would be to approximate the infinite series $\sum_{k=1}^{\infty} 1/k^2$ by adding the first few—or few hundred—terms. Unfortunately, because this series converges so slowly, such frontal attacks are not particularly illuminating. For instance,

$$1 + \frac{1}{4} + \frac{1}{9} + \cdots + \frac{1}{100} \approx 1.54977 \text{ (ten terms)};$$

$$1 + \frac{1}{4} + \frac{1}{9} + \cdots + \frac{1}{10000} \approx 1.63498 \text{ (one hundred terms)};$$

and

$$1 + \frac{1}{4} + \frac{1}{9} + \cdots + \frac{1}{1000^2} \approx 1.64393 \text{(one thousand terms)}.$$

We now know that, in spite of its prodigious number of terms, this last result is correct to only *two* decimal places. Other than the comforting fact that all these partial sums remain below 2.000 (as Bernoulli's comparison test had proved), direct numerical approximation is of little value.

Then, in a 1731 paper, the young Euler found a way to improve dramatically such numerical approximations. His discovery, remarkable in its vision and fearless in its manipulation of symbols, was truly ingenious.[4]

Euler's trick was to evaluate the (improper) integral

$$I = \int_0^{1/2} -\frac{\ln(1-t)}{t} \, dt$$

in two different ways. On the one hand, he replaced $\ln(1 - t)$ by its series expansion (see Chapter 2) and integrated termwise to get:

$$I = \int_0^{1/2} -\frac{-t - t^2/2 - t^3/3 - t^4/4 - \cdots}{t} \, dt$$

$$= \int_0^{1/2} \left(1 + \frac{t}{2} + \frac{t^2}{3} + \frac{t^3}{4} + \cdots\right) dt$$

$$= t + \frac{t^2}{4} + \frac{t^3}{9} + \frac{t^4}{16} + \cdots \Big|_0^{1/2}$$

---

[4]Euler, *Opera Omnia*, Ser. 1, Vol. 14, pp. 39–41.

$$= \frac{1}{2} + \frac{1/2^2}{4} + \frac{1/2^3}{9} + \frac{1/2^4}{16} + \cdots . \qquad (3.1)$$

On the other hand, he substituted $z = 1 - t$ to transform the original integral as follows:

$$I = \int_0^{1/2} -\frac{\ln(1-t)}{t} \, dt = \int_1^{1/2} \frac{\ln z}{1-z} \, dz$$

$$= \int_1^{1/2} (1 + z + z^2 + z^3 + \cdots) \ln z \, dz$$

$$= \int_1^{1/2} \ln z \, dz + \int_1^{1/2} z \ln z \, dz + \int_1^{1/2} z^2 \ln z \, dz + \int_1^{1/2} z^3 \ln z \, dz + \cdots,$$

because $1/(1-z)$ is the sum of the geometric series $1 + z + z^2 + z^3 + \cdots$.
Integration by parts implies that

$$\int_1^{1/2} z^n \ln z \, dz = \frac{z^{n+1}}{n+1} \ln z - \frac{z^{n+1}}{(n+1)^2} \Bigg|_1^{1/2},$$

and so this last expression becomes:

$$I = (z \ln z - z) + \left(\frac{z^2}{2} \ln z - \frac{z^2}{4} \right) + \left(\frac{z^3}{3} \ln z - \frac{z^3}{9} \right)$$

$$+ \left(\frac{z^4}{4} \ln z - \frac{z^4}{16} \right) + \cdots \Bigg|_1^{1/2}$$

$$= \ln z \left[z + \frac{z^2}{2} + \frac{z^3}{3} + \frac{z^4}{4} + \cdots \right] - \left(z + \frac{z^2}{4} + \frac{z^3}{9} + \frac{z^4}{16} + \cdots \right) \Bigg|_1^{1/2}$$

$$= \ln z \left[-\ln(1-z) \right] - \left(z + \frac{z^2}{4} + \frac{z^3}{9} + \frac{z^4}{16} + \cdots \right) \Bigg|_1^{1/2}$$

$$= - \left[\ln \left(\frac{1}{2} \right) \right]^2 - \left(\frac{1}{2} + \frac{1/2^2}{4} + \frac{1/2^3}{9} + \frac{1/2^4}{16} + \cdots \right)$$

$$+ [\ln 1][\ln 0] + \sum_{k=1}^{\infty} \frac{1}{k^2}.$$

Euler simply discarded $[\ln 1][\ln 0]$, although the modern reader might prefer to invoke l'Hôpital's Rule to verify that $\lim_{z \to 1^-} [\ln z][\ln(1-z)] = 0$. In any case, he arrived at:

$$I = -[\ln 2]^2 - \left(\frac{1}{2} + \frac{1/2^2}{4} + \frac{1/2^3}{9} + \frac{1/2^4}{16} + \cdots \right) + \sum_{k=1}^{\infty} \frac{1}{k^2}. \quad (3.2)$$

Then he equated the expressions for I in (3.1) and (3.2) and solved:

$$\sum_{k=1}^{\infty} \frac{1}{k^2} = 2 \left(\frac{1}{2} + \frac{1/2^2}{4} + \frac{1/2^3}{9} + \frac{1/2^4}{16} + \cdots \right) + [\ln 2]^2$$

$$= \sum_{k=1}^{\infty} \frac{1}{k^2 2^{k-1}} + [\ln 2]^2.$$

By this time the reader must have noticed a number of symbolic manipulations that require careful handling. Euler paid no heed to such matters as the existence of improper integrals or the termwise integration of infinite series. Nevertheless, his fusion of the log series, the geometric series, and integration by parts—all with the object of reaching an alternative expression for $\sum_{k=1}^{\infty} 1/k^2$—was a masterstroke. What made this effort worthwhile was that the resulting formula

$$\sum_{k=1}^{\infty} \frac{1}{k^2 2^{k-1}} + [\ln 2]^2$$

consists of a rapidly converging series (thanks to the 2^{k-1} term in the denominator) along with the number $[\ln 2]^2$, which Euler knew to dozens of decimal places. Using only fourteen terms of this new formula, one finds that $\sum_{k=1}^{\infty} 1/k^2 \approx 1.644934$, an answer correct to six places. This is far more accurate than summing a *thousand* terms of the original series. Euler's ingenuity had paid off.

Or had it? In spite of this vastly improved estimate, it was still just an estimate. Jakob Bernoulli, one remembers, had challenged the world to find the exact sum. In this sense, the problem seemed as far from resolution as ever.

But the end was in sight. Four years later, in 1735, Euler finally succeeded where so many others had failed. Admitting that his previous efforts had fallen short and that "it seemed most unlikely to be able to find anything new about this," Euler wrote with obvious joy:[5]

> Now, however, against all expectation I have found an elegant expression for the sum of the series $1 + \frac{1}{4} + \frac{1}{9} + \frac{1}{16} +$ etc., which depends on the quadrature of the circle... I have found that six times the sum

---

[5] Ibid., pp. 73–74.

of this series is equal to the square of the circumference of a circle whose diameter is 1.

To us, this wording about diameters and circumferences seems round-about—both geometrically and metaphorically—but because the circumference of such a circle has length π, Euler was asserting (in modern notation) that

$$\sum_{k=1}^{\infty} \frac{1}{k^2} = \frac{\pi^2}{6}.$$

Ever since, this has stood as one of the most wonderful formulas in mathematics. Those seeing it for the first time are puzzled by the unexpected appearance of π in a series of perfect squares, and at first glance it looks more like a typo than a theorem. Never fear: Euler was right.

His brief argument required two modest observations and one typically Eulerian leap of faith. First of all, we note that if $P(x) = 0$ is an nth degree polynomial equation with non-zero roots $a_1, a_2, a_3, \ldots, a_n$ and such that $P(0) = 1$, then in factored form

$$P(x) = \left(1 - \frac{x}{a_1}\right)\left(1 - \frac{x}{a_2}\right)\left(1 - \frac{x}{a_3}\right)\cdots\left(1 - \frac{x}{a_n}\right).$$

This is self-evident, because substituting $x = 0$ gives $P(0) = 1$, just as substituting $x = a_k$ yields $P(a_k) = 0$ for $k = 1, 2, \ldots n$.

Second, he needed the series expansion of $\sin x$, namely

$$\sin x = x - \frac{x^3}{3!} + \frac{x^5}{5!} - \frac{x^7}{7!} + \frac{x^9}{9!} - \cdots.$$

This formula, recognizable to any calculus student, was well known to Euler. (In Chapter 5, we shall discuss his derivation of this expansion, one whose use of the infinitely large and infinitely small is reminiscent of his development of the series for $\log(1 + x)$ from Chapter 2.)

These were the prerequisites underpinning his great discovery. The leap of faith was a belief that whatever holds for an ordinary polynomial will likewise hold for an "infinite polynomial." In this case, he assumed that a polynomial-like expression with infinitely many roots can be factored as $P(x)$ was factored above. Euler offered no proof of this, but for one who believed in the universality of formulas, it was a natural symbolic extension.

We now are ready for Euler's solution of the Basel problem.[6]

---

[6]Ibid., pp. 84–85.

Theorem. $\displaystyle\sum_{k=1}^{\infty} \frac{1}{k^2} = \frac{\pi^2}{6}$.

Proof. Euler introduced

$$P(x) = 1 - \frac{x^2}{3!} + \frac{x^4}{5!} - \frac{x^6}{7!} + \frac{x^8}{9!} - \cdots,$$

which he regarded as an "infinite polynomial." Clearly $P(0) = 1$. To find the roots of $P(x) = 0$, note that for $x \neq 0$,

$$P(x) = x \left[\frac{1 - x^2/3! + x^4/5! - x^6/7! + x^8/9! - \cdots}{x} \right]$$

$$= \frac{x - x^3/3! + x^5/5! - x^7/7! + x^9/9! - \cdots}{x} = \frac{\sin x}{x}.$$

So $P(x) = 0$ implies that $\sin x = 0$, which means in turn that $x = \pm k\pi$ for $k = 1, 2, \ldots$. Note that $x = 0$ is *not* a solution to $P(x) = 0$ because $P(0) = 1$.

In light of the observation above, he now factored $P(x)$ as:

$$1 - \frac{x^2}{3!} + \frac{x^4}{5!} - \frac{x^6}{7!} + \cdots = P(x)$$

$$= \left(1 - \frac{x}{\pi}\right)\left(1 - \frac{x}{-\pi}\right)\left(1 - \frac{x}{2\pi}\right)\left(1 - \frac{x}{-2\pi}\right)$$

$$\times \left(1 - \frac{x}{3\pi}\right)\left(1 - \frac{x}{-3\pi}\right) \cdots \qquad (3.3)$$

$$= \left[1 - \frac{x^2}{\pi^2}\right]\left[1 - \frac{x^2}{4\pi^2}\right]\left[1 - \frac{x^2}{9\pi^2}\right]\left[1 - \frac{x^2}{16\pi^2}\right] \cdots.$$

This is the chapter's most important formula. Euler had written $P(x)$ in two very different ways, equating the infinite sum on the left with the infinite product on the right.

What next? For Euler, nothing could be more natural than to expand the right side of (3.3) to get:

$$1 - \frac{x^2}{3!} + \frac{x^4}{5!} - \frac{x^6}{7!} + \frac{x^8}{9!} - \cdots$$

$$= 1 - \left(\frac{1}{\pi^2} + \frac{1}{4\pi^2} + \frac{1}{9\pi^2} + \frac{1}{16\pi^2} + \cdots\right)x^2 + \cdots \qquad (3.4)$$

where the coefficients of x^4 and higher (even) powers are unnecessary and, for the moment, unknown. He then equated the coefficients of x^2 in (3.4) to get

$$-\frac{1}{3!} = -\left(\frac{1}{\pi^2} + \frac{1}{4\pi^2} + \frac{1}{9\pi^2} + \frac{1}{16\pi^2} + \cdots\right)$$

$$= -\frac{1}{\pi^2}\left(1 + \frac{1}{4} + \frac{1}{9} + \frac{1}{16} + \cdots\right),$$

and concluded in dramatic fashion that

$$1 + \frac{1}{4} + \frac{1}{9} + \frac{1}{16} + \cdots = \frac{\pi^2}{6}. \qquad\qquad \text{Q.E.D.}$$

As he had promised, six times the sum of the series is the square of π. The Basel Problem was solved.

Of course Euler stands open to the charge of playing fast and loose with the logic. Over time, even *he* appeared troubled by the course his argument had taken and in later writings provided alternative—and what he considered more rigorous—derivations of this same formula. We shall examine one of these in the chapter's epilogue. Although none was entirely successful by modern standards, the reader should be assured that fully rigorous proofs have subsequently confirmed Euler's result.[7]

Such misgivings aside, Euler was confident that he had answered Bernoulli's unresolved question. There were internal indications that bolstered this certainty. For instance, a quick calculation revealed that $\pi^2/6 \approx 1.644934$, the precise estimate Euler had discovered a few years earlier. Numerically, he was right on target.

Moreover, his line of reasoning led to a previously known gem: Wallis's formula. In 1655, the English mathematician John Wallis (1616–1703), considering a different question and following a different logical path, had demonstrated that

$$\frac{2}{\pi} = \frac{1\cdot3\cdot3\cdot5\cdot5\cdot7\cdot7\cdot9\cdots}{2\cdot2\cdot4\cdot4\cdot6\cdot6\cdot8\cdot8\cdots}.$$

In the *Introductio*, Euler showed how the infinite product of (3.3) led to an alternate derivation of Wallis's formula. Putting $x = \pi/2$ into that expression yields

$$P\left(\frac{\pi}{2}\right) = \left[1 - \frac{(\pi/2)^2}{\pi^2}\right]\left[1 - \frac{(\pi/2)^2}{4\pi^2}\right]\left[1 - \frac{(\pi/2)^2}{9\pi^2}\right]\left[1 - \frac{(\pi/2)^2}{16\pi^2}\right]\cdots,$$

---

[7]Dan Kalman, "Six Ways to Sum a Series," *The College Mathematics Journal*, Vol. 24, No. 5, 1993, pp. 402–421.

which simplifies to

$$\frac{\sin(\pi/2)}{\pi/2} = \left[1 - \frac{1}{4}\right]\left[1 - \frac{1}{16}\right]\left[1 - \frac{1}{36}\right]\left[1 - \frac{1}{64}\right]\cdots$$

$$= \frac{3}{4} \times \frac{15}{16} \times \frac{35}{36} \times \cdots.$$

In short,

$$\frac{2}{\pi} = \frac{1 \cdot 3 \cdot 3 \cdot 5 \cdot 5 \cdot 7 \cdot 7 \cdot 9 \cdots}{2 \cdot 2 \cdot 4 \cdot 4 \cdot 6 \cdot 6 \cdot 8 \cdot 8 \cdots}.$$

Here we have Wallis's formula as a corollary. Surely this established that Euler's train of thought had not derailed. If his argument could recover previously known results such as this, there seemed all the more reason to embrace his initial conclusion.[8]

Quickly Euler's discovery flashed around the European mathematical community (if "flashed" is the correct verb to characterize eighteenth-century mail service). When Johann Bernoulli learned of the solution he wrote:

Utinam Frater superstes effet !
(If only my brother were alive!)[9]

André Weil called this "One of Euler's most sensational early discoveries, perhaps the one which established his growing reputation most firmly."[10] After this triumph, anyone who counted in European mathematics knew of the young genius who had succeeded so brilliantly where all others had failed.

It is easy to imagine that such success would lead many people to sit back, accept the plaudits of colleagues, and live off their well-deserved reputations. This was not Euler's way. On the contrary, once he had grasped a fruitful idea, he held on with an iron grip, squeezing out every last drop of information in an awesome exhibition of both genius and tenacity. So it was in this case.

For instance, he turned his attention to finding the exact sum of p-series with $p > 2$. Euler realized that this would require him to determine explicitly the coefficients of x^4, x^6, and so on in equation (3.4). Fortunately the tools necessary for such a determination were available in what are now called "Newton's formulas." These, published in Newton's *Arithmetica Universalis*,

---

[8] Euler, *Introduction to Analysis of the Infinite*, Book I, pp. 154–155.
[9] Johann Bernoulli, *Opera Omnia*, Vol. 4, Georg Olms Verlagsbuchhandlung, Hildesheim, 1968 (Reprint), p. 22.
[10] Weil, p. 184.

describe the links between the roots and the coefficients of a polynomial. In Newton's words:

> ... the coefficient of the second term in an equation is, if its sign be changed, equal to the aggregate of all the roots under their proper signs; that of the third equal to the aggregate of the products of the separate roots two at a time; that of the fourth, if its sign be changed, equal to the aggregate of the products of the individual roots three at a time; that of the fifth equal to the aggregate of the products of the roots four at a time; and so on indefinitely.[11]

Here we shall give Euler's derivation of formulas—equivalent to Newton's—relating the roots and coefficients.[12] His proof, which dates from 1750, took a most unusual path to his desired end, unexpectedly introducing techniques of differential calculus to solve a problem in algebra. Yet, promised Euler,

> even if [this derivation] seem exceedingly remote, nevertheless it perfectly resolves the entire situation.

His argument is so delightful, so thoroughly "Eulerian," that it deserves our attention.

Theorem. *If the nth degree polynomial* $P(y) = y^n - Ay^{n-1} + By^{n-2} - Cy^{n-3} + \cdots \pm N$ *is factored as* $P(y) = (y - r_1)(y - r_2) \cdots (y - r_n)$, *then*

$$\sum_{k=1}^{n} r_k = A,$$

$$\sum_{k=1}^{n} r_k^2 = A \sum_{k=1}^{n} r_k - 2B,$$

$$\sum_{k=1}^{n} r_k^3 = A \sum_{k=1}^{n} r_k^2 - B \sum_{k=1}^{n} r_k + 3C,$$

$$\sum_{k=1}^{n} r_k^4 = A \sum_{k=1}^{n} r_k^3 - B \sum_{k=1}^{n} r_k^2 + C \sum_{k=1}^{n} r_k - 4D, \text{ and so on.}$$

Proof. Euler's objective was to connect the polynomial's coefficients A, B, $C, \ldots, N$ and its roots $r_1, r_2, \ldots, r_n$. His first step, somewhat surprisingly, was

---

[11] Whiteside, ed., *The Mathematical Papers of Isaac Newton*, Vol. 5, p. 359.
[12] Euler, *Opera Omnia*, Ser. 1, Vol. 6, pp. 20–25.

to take logs:

$$\ln P(y) = \ln(y - r_1) + \ln(y - r_2) + \cdots + \ln(y - r_n).$$

The next step was more unanticipated—he differentiated both sides to get:

$$\frac{P'(y)}{P(y)} = \frac{1}{y - r_1} + \frac{1}{y - r_2} + \cdots + \frac{1}{y - r_n}. \tag{3.5}$$

As a final bit of analytic magic, Euler converted each fraction $1/(y - r_k)$ into its equivalent geometric series:

$$\frac{1}{y - r_k} = \frac{1}{y}\left(\frac{1}{1 - (r_k/y)}\right) = \frac{1}{y}\left(1 + \frac{r_k}{y} + \frac{r_k^2}{y^2} + \cdots\right)$$

$$= \frac{1}{y} + \frac{r_k}{y^2} + \frac{r_k^2}{y^3} + \frac{r_k^3}{y^4} + \cdots .$$

Therefore by (3.5)

$$\frac{P'(y)}{P(y)} = \frac{1}{y - r_1} + \frac{1}{y - r_2} + \cdots + \frac{1}{y - r_n}$$

$$= \frac{n}{y} + \left[\sum_{k=1}^{n} r_k\right]\frac{1}{y^2} + \left[\sum_{k=1}^{n} r_k^2\right]\frac{1}{y^3} + \left[\sum_{k=1}^{n} r_k^3\right]\frac{1}{y^4} + \cdots . \tag{3.6}$$

Note that this expresses $P'(y)/P(y)$ in terms of the *roots* of the original polynomial.

Because $P(y) = y^n - Ay^{n-1} + By^{n-2} - Cy^{n-3} + \cdots \pm N$, we have the obvious alternative

$$\frac{P'(y)}{P(y)} = \frac{ny^{n-1} - A(n-1)y^{n-2} + B(n-2)y^{n-3} - C(n-3)y^{n-4} + \cdots}{y^n - Ay^{n-1} + By^{n-2} - Cy^{n-3} + \cdots \pm N}, \tag{3.7}$$

framed in terms of the *coefficients* of the polynomial. Yet again, Euler had found different formulas for the same quantity, a ploy we have seen him use to good effect twice before in this chapter.

Equating the expressions from (3.6) and (3.7), he cross-multiplied to get:

$$ny^{n-1} - A(n-1)y^{n-2} + B(n-2)y^{n-3} - C(n-3)y^{n-4} + \cdots$$

$$= (y^n - Ay^{n-1} + By^{n-2} - Cy^{n-3} + \cdots \pm N)$$

$$\times \left(\frac{n}{y} + \left[\sum_{k=1}^{n} r_k \right] \frac{1}{y^2} + \left[\sum_{k=1}^{n} r_k^2 \right] \frac{1}{y^3} + \cdots \right)$$

$$= ny^{n-1} + \left(-nA + \sum_{k=1}^{n} r_k \right) y^{n-2}$$

$$+ \left(nB - A \sum_{k=1}^{n} r_k + \sum_{k=1}^{n} r_k^2 \right) y^{n-3} - \cdots.$$

Both sides of this equation begin with ny^{n-1}. Thereafter, we compare coefficients of like powers of y and solve to get the desired relationships. For example, equating the coefficients of y^{n-2} yields:

$$-A(n-1) = -nA + \sum_{k=1}^{n} r_k, \quad \text{and thus} \quad \sum_{k=1}^{n} r_k = A.$$

From the coefficients of y^{n-3}, we get:

$$B(n-2) = nB - A \sum_{k=1}^{n} r_k + \sum_{k=1}^{n} r_k^2, \quad \text{so that} \quad \sum_{k=1}^{n} r_k^2 = A \sum_{k=1}^{n} r_k - 2B.$$

Indeed, one can push this many terms deeper into the expansion (as Euler did) to find

$$\sum_{k=1}^{n} r_k^3 = A \sum_{k=1}^{n} r_k^2 - B \sum_{k=1}^{n} r_k + 3C$$

and

$$\sum_{k=1}^{n} r_k^4 = A \sum_{k=1}^{n} r_k^3 - B \sum_{k=1}^{n} r_k^2 + C \sum_{k=1}^{n} r_k - 4D,$$

and so on, with each new sum expressed in terms of previous ones. These are the promised relationships.   Q.E.D.

Convinced? There surely are points here deserving closer attention. For instance, when considering $\ln(y - r_k)$, Euler implicitly assumed that $y > r_k$. When expanding

$$\frac{1}{y - r_k} = \frac{1}{y} + \frac{r_k}{y^2} + \frac{r_k^2}{y^3} + \frac{r_k^3}{y^4} + \cdots$$

as a geometric series, an unspoken convergence assumption was present. Such matters become problematic should one extend these rules to an "infinite polynomial," which is exactly what Euler did.

Still, it is impossible not to be struck again by Euler's brilliance in attacking an algebraic theorem about roots and coefficients by means of logarithms, derivatives, and geometric series—all tools from his analytic arsenal. His was an extremely agile mind.

What do these formulas have to do with summing p-series? To answer that question, we consider a polynomial containing only even powers of x and factored as follows:

$$1 - Ax^2 + Bx^4 - Cx^6 + \cdots \pm Nx^{2n} = (1 - r_1 x^2)(1 - r_2 x^2) \cdots (1 - r_n x^2).$$

$$(3.8)$$

Substitute $1/y$ for x^2:

$$1 - A\left(\frac{1}{y}\right) + B\left(\frac{1}{y}\right)^2 - C\left(\frac{1}{y}\right)^3 + \cdots \pm N\left(\frac{1}{y}\right)^n$$
$$= \left(1 - r_1\frac{1}{y}\right)\left(1 - r_2\frac{1}{y}\right) \cdots \left(1 - r_n\frac{1}{y}\right).$$

Then multiply both sides by y^n to get:

$$y^n - Ay^{n-1} + By^{n-2} - Cy^{n-3} + \cdots \pm N = (y - r_1)(y - r_2) \cdots (y - r_n).$$

This of course is precisely the case Euler considered above. Hence for (3.8) also we have the formulas

(a) $\displaystyle\sum_{k=1}^{n} r_k = A,$

(b) $\displaystyle\sum_{k=1}^{n} r_k^2 = A\sum_{k=1}^{n} r_k - 2B,$ and

(c) $\displaystyle\sum_{k=1}^{n} r_k^3 = A\sum_{k=1}^{n} r_k^2 - B\sum_{k=1}^{n} r_k + 3C.$

Euler assumed that these relationships between coefficients and roots remain valid *even if both are infinitely plentiful*—that is, when the sum runs from $k = 1$ to ∞. He returned to (3.3)

$$1 - \frac{x^2}{3!} + \frac{x^4}{5!} - \frac{x^6}{7!} + \frac{x^8}{9!} - \cdots$$

$$= \left[1 - \frac{x^2}{\pi^2}\right]\left[1 - \frac{x^2}{4\pi^2}\right]\left[1 - \frac{x^2}{9\pi^2}\right]\left[1 - \frac{x^2}{16\pi^2}\right]\cdots,$$

which looks exactly like an infinite version of (3.8) with $A = 1/3!$, $B = 1/5!$, $C = 1/7!$, and $r_k = 1/k^2\pi^2$ for $k = 1, 2, \ldots$.

According to (a), $\sum_{k=1}^{\infty} 1/k^2\pi^2 = 1/3! = 1/6$ and so $\sum_{k=1}^{\infty} 1/k^2 = \pi^2/6$. This, of course, is Euler's "sensational" result derived above. But (b) and (c) yield entirely new information:

(b) $\displaystyle\sum_{k=1}^{\infty}\left(\frac{1}{k^2\pi^2}\right)^2 = A\sum_{k=1}^{\infty}\frac{1}{k^2\pi^2} - 2B = \left(\frac{1}{3!}\right)^2 - \frac{2}{5!} = \frac{1}{90}$,

and so $\displaystyle\sum_{k=1}^{\infty}\frac{1}{k^4} = \frac{\pi^4}{90}$;

(c) $\displaystyle\sum_{k=1}^{\infty}\left(\frac{1}{k^2\pi^2}\right)^3 = A\sum_{k=1}^{\infty}\left(\frac{1}{k^2\pi^2}\right)^2 - B\sum_{k=1}^{\infty}\frac{1}{k^2\pi^2} + 3C$

$$= \left(\frac{1}{3!}\right)\left(\frac{1}{90}\right) - \left(\frac{1}{5!}\right)\left(\frac{1}{6}\right) + 3\left(\frac{1}{7!}\right) = \frac{1}{945},$$

and thus $\displaystyle\sum_{k=1}^{\infty}\frac{1}{k^6} = \frac{\pi^6}{945}$.

These are very strange. In his original paper Euler pushed further to evaluate p-series for $p = 8$, 10, and 12 . Later, in a 1744 publication, he gave exact sums for even values of p up to the colossal, if slightly ridiculous,[13]

$$\sum_{k=1}^{\infty}\frac{1}{k^{26}} = \frac{2^{24}}{27!}(76977927\pi^{26}) = \frac{1315862}{11094481976030578125}\pi^{26}.$$

Here Euler was answering questions no one had ever before *asked*. Better yet, his work contained the seeds for future research, including a link to what are now called the Bernoulli numbers and a hint of the Riemann zeta function that would prove so significant in the nineteenth century.[14] It was indeed an impressive display by a young mathematician aptly described by François Arago as "analysis incarnate."[15]

---

[13] Euler, *Opera Omnia*, Ser. 1, Vol. 14, p. 185.

[14] Ayoub, pp. 1067–1086.

[15] Howard Eves, *An Introduction to the History of Mathematics*, 5th ed., Saunders, New York, 1983, p. 330.

Epilogue

Here we shall address three topics related to the work of this chapter. First, we provide Euler's alternate solution of the Basel Problem. Second, we describe his *application* of the discoveries recounted above. And finally, we discuss a subsidiary challenge that has resisted the efforts of Euler and all who followed.

As noted, some of Euler's contemporaries, while accepting his answer to the Basel Problem, wondered about the validity of the argument that got him there. Daniel Bernoulli was especially concerned and wrote Euler in this regard.[16] In an attempt to silence such doubters, Euler devised another, quite different, proof that $\sum_{k=1}^{\infty} 1/k^2 = \pi^2/6$. Although unlike the first, it is every bit as masterful.[17]

This argument requires three preliminary results, each of which falls well within the scope of a modern calculus course.

A. Prove the identity $\dfrac{1}{2}(\sin^{-1} x)^2 = \displaystyle\int_0^x \dfrac{\sin^{-1} t}{\sqrt{1-t^2}}\, dt$:

This follows immediately from the substitution $u = \sin^{-1} t$.

B. Find a series expansion for $\sin^{-1} x$:

Recalling that

$$\sin^{-1} x = \int_0^x \frac{1}{\sqrt{1-t^2}}\, dt = \int_0^x (1-t^2)^{-1/2}\, dt,$$

we replace the expression under the integral by its binomial series and integrate termwise to get

$$\sin^{-1} x = \int_0^x \left(1 + \frac{1}{2}t^2 + \frac{1\cdot 3}{2^2\cdot 2!}t^4 + \frac{1\cdot 3\cdot 5}{2^3\cdot 3!}t^6 + \frac{1\cdot 3\cdot 5\cdot 7}{2^4\cdot 4!}t^8 + \cdots \right) dt$$

$$= t + \frac{1}{2}\times\frac{t^3}{3} + \frac{1\cdot 3}{2\cdot 4}\times\frac{t^5}{5} + \frac{1\cdot 3\cdot 5}{2\cdot 4\cdot 6}\times\frac{t^7}{7}$$

$$+ \frac{1\cdot 3\cdot 5\cdot 7}{2\cdot 4\cdot 6\cdot 8}\times\frac{t^9}{9} + \cdots \Bigg|_0^x$$

---

[16] Euler, *Opera Omnia*, Ser. 1, Vol. 14, p. 141.
[17] Ibid., pp. 178–181.

$$= x + \frac{1}{2} \times \frac{x^3}{3} + \frac{1 \cdot 3}{2 \cdot 4} \times \frac{x^5}{5} + \frac{1 \cdot 3 \cdot 5}{2 \cdot 4 \cdot 6} \times \frac{x^7}{7}$$

$$+ \frac{1 \cdot 3 \cdot 5 \cdot 7}{2 \cdot 4 \cdot 6 \cdot 8} \times \frac{x^9}{9} + \cdots .$$

C. Prove the relation $\displaystyle\int_0^1 \frac{t^{n+2}}{\sqrt{1-t^2}}\, dt = \frac{n+1}{n+2} \int_0^1 \frac{t^n}{\sqrt{1-t^2}}\, dt$ for $n \geq 1$:

For

$$J = \int_0^1 \frac{t^{n+2}}{\sqrt{1-t^2}}\, dt,$$

apply integration by parts with $u = t^{n+1}$ and $dv = (t/\sqrt{1-t^2})\, dt$ to get

$$J = \left. (-t^{n+1}\sqrt{1-t^2}) \right|_0^1 + (n+1)\int_0^1 t^n \sqrt{1-t^2}\, dt$$

$$= 0 + (n+1)\int_0^1 \frac{t^n(1-t^2)}{\sqrt{1-t^2}}\, dt = (n+1)\int_0^1 \frac{t^n}{\sqrt{1-t^2}}\, dt - (n+1)J.$$

Therefore

$$(n+2)J = (n+1)\int_0^1 \frac{t^n}{\sqrt{1-t^2}}\, dt,$$

and the result follows.

Fine. We now follow Euler in assembling these components to re-prove his formula. Simply let $x = 1$ in (A) to get:

$$\frac{\pi^2}{8} = \frac{1}{2}(\sin^{-1} 1)^2 = \int_0^1 \frac{\sin^{-1} t}{\sqrt{1-t^2}}\, dt.$$

Next, replace $\sin^{-1} t$ with its series expansion from (B) and integrate termwise:

$$\frac{\pi^2}{8} = \int_0^1 \frac{t}{\sqrt{1-t^2}}\, dt + \frac{1}{2 \cdot 3}\int_0^1 \frac{t^3}{\sqrt{1-t^2}}\, dt + \frac{1 \cdot 3}{2 \cdot 4 \cdot 5}\int_0^1 \frac{t^5}{\sqrt{1-t^2}}\, dt$$

$$+ \frac{1 \cdot 3 \cdot 5}{2 \cdot 4 \cdot 6 \cdot 7}\int_0^1 \frac{t^7}{\sqrt{1-t^2}}\, dt + \cdots .$$

Knowing that

$$\int_0^1 \frac{t}{\sqrt{1-t^2}}\, dt = 1,$$

we evaluate the other integrals using the recursion in (C):

$$\frac{\pi^2}{8} = 1 + \frac{1}{2 \cdot 3}\left[\frac{2}{3}\right] + \frac{1 \cdot 3}{2 \cdot 4 \cdot 5}\left[\frac{2}{3} \times \frac{4}{5}\right] + \frac{1 \cdot 3 \cdot 5}{2 \cdot 4 \cdot 6 \cdot 7}\left[\frac{2}{3} \times \frac{4}{5} \times \frac{6}{7}\right] + \cdots$$

$$= 1 + \frac{1}{9} + \frac{1}{25} + \frac{1}{49} + \cdots,$$

a summation involving only the odd squares.

From here Euler needed the following simple observation to reach his desired end.

Theorem. $\displaystyle\sum_{k=1}^{\infty} \frac{1}{k^2} = \frac{\pi^2}{6}.$

Proof.

$$\sum_{k=1}^{\infty} \frac{1}{k^2} = \left[1 + \frac{1}{9} + \frac{1}{25} + \frac{1}{49} + \cdots\right] + \left[\frac{1}{4} + \frac{1}{16} + \frac{1}{36} + \frac{1}{64} + \cdots\right]$$

$$= \left[1 + \frac{1}{9} + \frac{1}{25} + \frac{1}{49} + \cdots\right] + \frac{1}{4}\left[1 + \frac{1}{4} + \frac{1}{9} + \frac{1}{16} + \frac{1}{25} + \cdots\right]$$

$$= \frac{\pi^2}{8} + \frac{1}{4}\sum_{k=1}^{\infty} \frac{1}{k^2}.$$

Thus

$$\frac{3}{4}\sum_{k=1}^{\infty} \frac{1}{k^2} = \frac{\pi^2}{8},$$

and so

$$\sum_{k=1}^{\infty} \frac{1}{k^2} = \frac{4}{3} \times \frac{\pi^2}{8} = \frac{\pi^2}{6}. \qquad \text{Q.E.D.}$$

There, before us, is the solution of the Basel Problem. This derivation, so different from the first, is the work of an analyst at the top of his powers—and one who seems to be enjoying himself immensely.

The epilogue's second objective is to show Euler applying his formulas to other, seemingly unrelated, matters. Indeed, he asserted that the "principal use" of these results "is in the calculation of logarithms."[18] Although this claim may sound far-fetched, he was happy to explain what he had in mind.

---

[18] Euler, *Introduction to Analysis of the Infinite*, Book I, p. 158.

Consider again the chapter's pivotal equation, labeled (3.3):

$$P(x) = \left[1 - \frac{x^2}{\pi^2}\right]\left[1 - \frac{x^2}{4\pi^2}\right]\left[1 - \frac{x^2}{9\pi^2}\right]\left[1 - \frac{x^2}{16\pi^2}\right]\cdots.$$

Recalling that $P(x) = (\sin x)/x$ for $x \neq 0$, we cross-multiply to get the infinite product

$$\sin x = x\left[1 - \frac{x^2}{\pi^2}\right]\left[1 - \frac{x^2}{4\pi^2}\right]\left[1 - \frac{x^2}{9\pi^2}\right]\left[1 - \frac{x^2}{16\pi^2}\right]\cdots,$$

a result that holds even if $x = 0$.

When confronting this (or any) product, Euler seemed unable to resist taking logarithms. Such was the case here, and as usual it paid off:

$$\ln(\sin x) = \ln x + \ln\left(1 - \frac{x^2}{\pi^2}\right) + \ln\left(1 - \frac{x^2}{4\pi^2}\right) + \ln\left(1 - \frac{x^2}{9\pi^2}\right) + \cdots$$

which, for $x = \pi/n$, becomes

$$\ln\left(\sin\frac{\pi}{n}\right) = \ln\pi - \ln n + \ln\left(1 - \frac{1}{n^2}\right)$$

$$+ \ln\left(1 - \frac{1}{4n^2}\right) + \ln\left(1 - \frac{1}{9n^2}\right) + \cdots.$$

Perhaps the reader is by now sufficiently familiar with Euler's methods to anticipate that his next step was to introduce the series expansion of $\ln(1 - x)$ to get:

$$\ln\left(\sin\frac{\pi}{n}\right) = \ln\pi - \ln n + \left[-\frac{1}{n^2} - \frac{1}{2n^4} - \frac{1}{3n^6}\cdots\right]$$

$$+ \left[-\frac{1}{4n^2} - \frac{1}{32n^4} - \frac{1}{192n^6}\cdots\right]$$

$$+ \left[-\frac{1}{9n^2} - \frac{1}{162n^4} - \frac{1}{2187n^6}\cdots\right] + \cdots$$

$$= \ln\pi - \ln n - \frac{1}{n^2}\left(1 + \frac{1}{4} + \frac{1}{9} + \cdots\right)$$

$$- \frac{1}{2n^4}\left(1 + \frac{1}{16} + \frac{1}{81} + \cdots\right)$$

$$- \frac{1}{3n^6}\left(1 + \frac{1}{64} + \frac{1}{729} + \cdots\right) - \cdots.$$

Remarkably, this formula contains *precisely* the p-series Euler had evaluated. It follows that

$$\ln\left(\sin\frac{\pi}{n}\right) = \ln\pi - \ln n - \frac{1}{n^2}\left(\frac{\pi^2}{6}\right) - \frac{1}{2n^4}\left(\frac{\pi^4}{90}\right) - \frac{1}{3n^6}\left(\frac{\pi^6}{945}\right) - \cdots.$$

What emerges is a rapidly converging series for $\ln(\sin\pi/n)$. To see it in action, choose $n = 7$ and approximate

$$\ln\left(\sin\frac{\pi}{7}\right) = \ln\pi - \ln 7 - \frac{1}{49}\left(\frac{\pi^2}{6}\right) - \frac{1}{4802}\left(\frac{\pi^4}{90}\right)$$

$$-\frac{1}{352947}\left(\frac{\pi^6}{945}\right) - \cdots$$

$$\approx -0.83498,$$

which, with only five terms, is accurate to within ± 0.00000005.

Euler had found a way of computing logarithms of sines with great efficiency. More remarkably, he did so while short-cutting the numerical values of the sines themselves, as he himself observed when he wrote:

> [w]ith these formulas, we can find both the natural and the common logarithms of the sine and cosine of any angle, *even without knowing the sines and cosines.* [italics added][19]

In spite of such success, Euler got nowhere on a fundamental problem: to evaluate the p-series for odd values of p. Even the simplest of these,

$$\sum_{k=1}^{\infty}\frac{1}{k^3} = 1 + \frac{1}{8} + \frac{1}{27} + \frac{1}{64} + \frac{1}{125} + \frac{1}{216} + \frac{1}{343} + \cdots,$$

resisted explicit solution. Euler's original proof—as it emerged from equation (3.3)—was obviously geared toward even powers of x, and thus even values of p. Odd exponents slipped through his net.

Euler was keenly aware of the situation. The best he could do in his 1735 paper was to evaluate the loosely related series[20]

$$1 - \frac{1}{27} + \frac{1}{125} - \frac{1}{343} + \cdots = \sum_{k=0}^{\infty}(-1)^k\frac{1}{(2k+1)^3} = \frac{\pi^3}{32}.$$

This was an intriguing answer. Unfortunately, it was to the wrong question.

---

[19] Ibid., p. 165.
[20] Euler, *Opera Omnia*, Ser. 1, Vol. 14, p. 80.

For guidance, Euler again turned to numerical approximations.[21] Because $\sum_{k=1}^{\infty} 1/k^2 = \pi^2/6$ and $\sum_{k=1}^{\infty} 1/k^4 = \pi^4/90$, he naturally conjectured that $\sum_{k=1}^{\infty} 1/k^3 = \pi^3/m$ for some integer m falling between 6 and 90. With customary zeal, Euler calculated $\sum_{k=1}^{\infty} 1/k^3 \approx 1.202056903$ and, setting this equal to π^3/m, deduced that $m = 25.79435$—hardly a promising result.

At a later point, Euler conjectured that

$$\sum_{k=1}^{\infty} \frac{1}{k^3} = \alpha(\ln 2)^2 + \beta\frac{\pi^2}{6}\ln 2$$

for rational numbers α and β.[22] Intriguing though this was, it too led him nowhere.

So what do we know today about $\sum_{k=1}^{\infty} 1/k^3$? The answer is, "Disappointingly little." Progress over the centuries has been minimal. Indeed, only in 1978 did Roger Apéry manage to show that $\sum_{k=1}^{\infty} 1/k^3$ sums to an *irrational* number.[23] His was an ingenious answer to a difficult question. Yet the conclusion was both unsurprising and unsatisfying—unsurprising because the irrationality of this sum had been universally anticipated even if never proved; unsatisfying because one would have preferred an *exact* answer, not a broad classification like "irrational." It is as though we were looking for Captain Kidd's treasure and Apéry brilliantly demonstrated that it could be found somewhere in the Solar System. Mathematicians had wanted something a little more specific.

Worse, the irrationality of the series with $p = 3$ has as yet no counterpart for $p = 5$, $p = 7$, or any of the higher odd powers. For these, we are no further along than when Euler put down his pen over two centuries ago.

In this sense, even after 300 years, Jakob Bernoulli's problem is with us still. Faced with the mystery of the odd-valued p-series, one is tempted to throw up one's hands and reissue Jakob's challenge from 1689: "If anyone finds and communicates to us that which has thus far eluded our efforts, great will be our gratitude."

Then hope for a 21st century Euler.

---

[21] Ibid., p. 440.

[22] Euler, Opera Omnia, Ser. 1, Vol. 4, pp. 143–144.

[23] Alfred van der Poorten, "A Proof that Euler Missed," *The Mathematical Intelligencer*, Vol. 1, No. 4, 1978, pp. 195–203.

CHAPTER 4

Euler and Analytic Number Theory

Oft expectation fails, and most oft there
Where most it promises; and oft it hits
Where hope is coldest, and despair most sits.
— *All's Well that Ends Well*

Contrary to Shakespeare's opinion, our expectations usually *are* confirmed. People have a pretty accurate sense of what will succeed and what will fail. Be it a lead balloon, a bull in a china shop, or an onion milk shake, there are combinations that are instantly seen to be unworkable.

Yet on rare occasions an unlikely juxtaposition—where "hope is coldest"—proves surprisingly fruitful. This is true in life, as Shakespeare reminded us, and it is true in mathematics. Indeed, certain subdisciplines were created by joining two apparently unrelated branches of the subject—like algebraic topology, combinatorial algebra, or that most significant of all, analytic geometry.

It could be argued, however, that the most *unexpected* juxtaposition, the one that seems most unnatural, is analytic number theory. This branch of mathematics applies the techniques of calculus/analysis to the realm of the whole numbers. What makes it so peculiar is that analysis treats continuous, "flowing" phenomena. Its major tools—convergence and divergence, derivative and integral—require the rich continuum of the real number system. Number theory, by contrast, is as discrete as it gets. By no stretch of the imagination does one whole number flow into another. The integers are separated, isolated entities which require a very different set of tools.

These are strange bedfellows. The fusion of analysis and number theory seems to be the mathematical equivalent of that onion milk shake. Only a fool would waste time on such a combination.

Only a fool ... or a genius.

In fact, analytic number theory stands as one of the jewels in the mathematical crown. Difficult and profound, it is a subject that gained its voice in the nineteenth century but whose early murmurings can be traced to Euler, that most insightful scholar of the eighteenth.

Prologue

We touched upon (classical) number theory in Chapter 1. Even a novice quickly perceives the central role played by the primes. Because all positive integers greater than 1 can be uniquely factored into primes, these serve as the fundamental components—the bricks and mortar—of number theory. Learn about primes and you have gone a long way toward learning about whole numbers generally.

Over two millennia ago Euclid asked, and answered, a basic question about the primes in what is one of the greatest proofs ever devised. As Proposition 20 of Book IX of the *Elements*, he showed that no finite collection of primes—no matter how vast—can possibly include them all. Although his argument has been reproduced countless times, it always warrants a quick reprise.

Theorem. *No finite collection of primes includes them all.*

Proof. Let $p_1, p_2, \ldots, p_n$ be any finite set of primes. The objective is to prove there is a prime not included among them. To this end, let $M = (p_1 \times p_2 \times \cdots \times p_n) + 1$ and consider the two alternatives:

Case 1. If M is prime, then it surely is a "new" prime not in the original set, for it is larger than any of $p_1, p_2, \ldots, p_n$.

Case 2. If M is composite, then it has a prime factor q. We assert that q is not one of the original primes. For if $q = p_k$ for some k, then q would divide evenly into both M and $p_1 \times p_2 \times \cdots \times p_n$ and hence into their difference $M - p_1 \times p_2 \times \cdots \times p_n = 1$. But the prime q, being at least 2, cannot divide evenly into 1. This contradiction means that q, differing from all the p_k, is the new prime we sought.

By Cases 1 and 2, it is clear that any finite set of primes can be augmented. In short, there are infinitely many primes.   Q.E.D.

The reader is advised to savor this proof, for there is none more elegant in all of mathematics.

Euclid left the matter there, but prime numbers—their characteristics, their structure, their distribution—have been among the most studied of mathematical objects, exhibiting a fascination as endless as the primes themselves.

For instance, consider this dichotomy among the odd primes: each is either of the form $4k + 1$ or $4k - 1$. That is, an odd prime (indeed, any odd number) is either one more or one less than a multiple of 4.

Initially we might try to assess the relative abundance of the two types of prime. Among the first hundred numbers, the primes in each family are:

$$4k + 1 : \quad 5, 13, 17, 29, 37, 41, 53, 61, 73, 89, 97$$

$$4k - 1 : \quad 3, 7, 11, 19, 23, 31, 43, 47, 59, 67, 71, 79, 83.$$

Among the next hundred, we have:

$$4k + 1 : \quad 101, 109, 113, 137, 149, 157, 173, 181, 193, 197$$

$$4k - 1 : \quad 103, 107, 127, 131, 139, 151, 163, 167, 179, 191, 199.$$

Patterns here are not conspicuous, but from these scanty data one is tempted to propose that, in any string of whole numbers starting from 1, there are slightly more $4k - 1$ primes than their $4k + 1$ counterparts. In other words, one might surmise that, no matter the size of n, the $4k - 1$ primes are in the majority among the numbers $1, 2, 3, \ldots , n$.

This conjecture is false. The $4k + 1$ primes eventually overtake the $4k - 1$ primes, but (strangely) this does not happen until we consider the very long string $1, 2, 3, \ldots , 26861$. Only then does the balance finally tip in favor of the $4k + 1$ type. Soon thereafter it tips back—and then back again. The twentieth-century mathematician J. E. Littlewood (1885–1977) showed that the majority status changes hands infinitely often as we march through the positive integers.[1] It is like a two-horse race in which neither thoroughbred can maintain the lead.

Long before Littlewood, mathematicians had raised the question of the *overall* abundance of the two types of primes. Because there are infinitely many primes, it is clear that at least one of the two families must itself be infinite. With an argument modeled upon Euclid's, we prove:

Theorem. *There are infinitely many* $4k - 1$ *primes.*

Proof. To begin, observe that the product of $4k + 1$ primes—although certainly not prime—is another number of the form $4k + 1$. We demonstrate this for the product of two such numbers:

$$(4r + 1)(4s + 1) = 16rs + 4r + 4s + 1 = 4(4rs + r + s) + 1,$$

---

[1]David Wells, *The Penguin Dictionary of Curious and Interesting Numbers*, Penguin, New York, 1986, p. 176.

which is clearly one more than a multiple of 4. The result for arbitrary products follows by induction.

Now, assume we have a finite collection of $4k - 1$ primes, namely $p_1 = 4k_1 - 1, p_2 = 4k_2 - 1, \ldots$, and $p_n = 4k_n - 1$. In our quest for yet another prime of this form, we introduce $M = 4(p_1 \times p_2 \times \cdots \times p_n) - 1$.

Case 1. If M is prime, we are done, for M is surely larger than $p_1, p_2, \ldots$, or p_n and thus is a "new" $4k - 1$ prime.

Case 2. If M is composite, then one of its prime divisors *must* be of the form $4k - 1$. This follows because, if all the prime factors of M are of the $4k + 1$ type, then their product—namely M itself—would have to be of this type as well, as noted in our observation above. Such, of course, is not the case.

Therefore, M has at least one prime factor q of the form $4k - 1$. But if $q = p_i$ for some i, then q would evenly divide into both M and $4(p_1 \times p_2 \times \cdots \times p_n)$ and so would evenly divide the difference $4(p_1 \times p_2 \times \cdots \times p_n) - M = 1$. Again this is a contradiction because $q \geq 3$. Therefore, q is not only a $4k - 1$ prime but is a *different* one than $p_1, p_2, \ldots$, or p_n.

By Cases 1 and 2, we see that any finite collection of $4k - 1$ primes cannot contain all such primes. Thus there are infinitely many primes of this type.

Q.E.D.

And what of the abundance of the $4k + 1$ primes? The preceding theorem does not allow us to conclude anything in this direction. True, we know there are infinitely many primes altogether, and infinitely many of them are of the $4k - 1$ variety, but this is logically insufficient to determine the finitude or infinitude of the other type. The fact that there are infinitely many of the $4k + 1$ primes—although true—turned out to be much more difficult to prove. We shall see what Euler had to say on this subject later in the chapter.

In terms of abundance and relative distribution, the two kinds of odd primes seem essentially equivalent. Perhaps this is as one would guess. But there is a fundamental manner in which these families differ. It was Fermat who first conjectured, and Euler who first proved, the amazing proposition:

> A $4k + 1$ prime can be written as the sum of two perfect squares in one and only one way, whereas a $4k - 1$ prime cannot be written as the sum of two perfect squares in any way at all.

The reader is invited to seek such decompositions among the $4k + 1$ primes. For instance, $37 = 1 + 36 = 1^2 + 6^2$; $137 = 16 + 121 = 4^2 + 11^2$; $281 = 25 + 256 = 5^2 + 16^2$. Moreover, these decompositions into squares are unique. On the other hand, because the sum of two perfect squares can never be one less than a multiple of 4, primes of the second kind cannot be decomposed into two squares. This strange property, establishing a non-intuitive and quite spectacular difference between the two types of primes, is one of the great theorems of mathematics.

The results discussed thus far fall under the heading of "classical" number theory. They address properties of primality and divisibility but remain entirely within the discrete realm. All this would change with the surprising recognition that analytic techniques could be put to use in the study of whole numbers.

Enter Euler

Like all number theorists, Euler was intrigued by primes. Recall from Chapter 1 that it was he who disproved Fermat's conjecture that all numbers of the form $2^{2^n} + 1$ are prime. As we just mentioned, it was Euler who first proved the great dichotomy between the two classes of odd primes with regard to their decomposition into sums of squares. And who but a lover of primes would publish a paper with the awesome title "On a table of prime numbers up to a million and beyond" as Euler did in 1774?[2]

In 1737, however, his investigations crossed the line from "pure" number theory to a bolder, more analytic variety. We shall examine here a landmark paper, "*Variae observationes circa series infinitas*," where Euler was engaged in one of his favorite pastimes: summing infinite series.[3]

He began the paper by investigating some highly irregular series that had little to do with number theory. For instance, he proposed to find the exact sum of the infinite series

$$\frac{1}{15} + \frac{1}{63} + \frac{1}{80} + \frac{1}{255} + \frac{1}{624} + \cdots .$$

The pattern here is anything but evident. It is a perceptive reader who can figure out the next term—let alone determine the sum of the series itself.

Euler explained. He observed that the terms in the series are those reciprocals "whose denominators are one less than all perfect squares which simultaneously are other powers." As a case in point, $16 = 4^2 = 2^4$, and so 16 is a perfect square that is also a fourth power; hence $16 - 1 = 15$ qualifies as a denominator of a term in the series. So does $64 - 1 = 63$ because the perfect square $64(= 8^2 = 4^3)$ is also a perfect cube. But the perfect square 36 cannot

---

[2]Euler, *Opera Omnia*, Ser. 1, Vol. 3, pp. 359–404.
[3]Euler, *Opera Omnia*, Ser. 1, Vol. 14, pp. 216–244.

be written as *another* integer power; hence $\frac{1}{35}$ does not appear in this series. (The next term, by the way, is $\frac{1}{728}$.)

But what is the sum? Euler began with his famous result from Chapter 3,

$$\frac{\pi^2}{6} = 1 + \frac{1}{4} + \frac{1}{9} + \frac{1}{16} + \frac{1}{25} + \frac{1}{36} + \frac{1}{49} + \frac{1}{64} + \frac{1}{81} + \cdots,$$

which he regrouped as follows:

$$\frac{\pi^2}{6} - 1 = \left(\frac{1}{4} + \frac{1}{16} + \frac{1}{64} + \cdots\right) + \left(\frac{1}{9} + \frac{1}{81} + \frac{1}{729} + \cdots\right)$$

$$+ \left(\frac{1}{25} + \frac{1}{625} + \cdots\right) + \left(\frac{1}{36} + \frac{1}{1296} + \cdots\right) \tag{4.1}$$

$$+ \left(\frac{1}{49} + \frac{1}{2401} + \cdots\right) + \left(\frac{1}{100} + \frac{1}{10000} + \cdots\right) + \cdots.$$

He then evaluated the geometric series within the various parentheses to get:

$$\frac{\pi^2}{6} - 1 = \frac{1}{3} + \frac{1}{8} + \frac{1}{24} + \frac{1}{35} + \frac{1}{48} + \frac{1}{99} + \cdots.$$

Notice—and this is the critical observation—that each denominator here is one less than a perfect square that cannot otherwise be written as a power. Those squares that are also higher powers, like $\frac{1}{16}$ and $\frac{1}{64}$, are embedded within one of the geometric series in (4.1).

Euler could easily sum the reciprocals of *all* numbers falling one unit below perfect squares:

$$\frac{1}{3} + \frac{1}{8} + \frac{1}{15} + \frac{1}{24} + \frac{1}{35} + \frac{1}{48} + \frac{1}{63} + \frac{1}{80} + \frac{1}{99} + \cdots$$

$$= \frac{1}{2}\left[\frac{2}{3} + \frac{2}{8} + \frac{2}{15} + \frac{2}{24} + \frac{2}{35} + \frac{2}{48} + \frac{2}{63} + \frac{2}{80} + \frac{2}{99} + \cdots\right]$$

$$= \frac{1}{2}\left[\left(1 - \frac{1}{3}\right) + \left(\frac{1}{2} - \frac{1}{4}\right) + \left(\frac{1}{3} - \frac{1}{5}\right) + \left(\frac{1}{4} - \frac{1}{6}\right) + \left(\frac{1}{5} - \frac{1}{7}\right)\right.$$

$$\left. + \left(\frac{1}{6} - \frac{1}{8}\right) + \left(\frac{1}{7} - \frac{1}{9}\right) + \cdots\right]$$

$$= \frac{1}{2}\left[1 + \frac{1}{2}\right] = \frac{3}{4},$$

because the series telescopes.

To finish the problem, Euler needed only to subtract:

$$\frac{1}{15} + \frac{1}{63} + \frac{1}{80} + \frac{1}{255} + \frac{1}{624} + \frac{1}{728} + \cdots$$

$$= \left[\frac{1}{3} + \frac{1}{8} + \frac{1}{15} + \frac{1}{24} + \frac{1}{35} + \frac{1}{48} + \frac{1}{63} + \cdots \right]$$

$$- \left[\frac{1}{3} + \frac{1}{8} + \frac{1}{24} + \frac{1}{35} + \frac{1}{48} + \frac{1}{99} + \cdots \right]$$

$$= \frac{3}{4} - \left[\frac{\pi^2}{6} - 1 \right] = \frac{7}{4} - \frac{\pi^2}{6},$$

a result as weird as anyone could wish. This is vintage Euler, manipulating formulas with glee.

A few pages deeper into the paper, he attacked the harmonic series with the same sort of gusto. Of course, he knew that the harmonic series diverged to infinity, but such knowledge hardly stopped him. He deduced a strange and wonderful connection between the harmonic series and the prime numbers by showing that

$$1 + \frac{1}{2} + \frac{1}{3} + \frac{1}{4} + \frac{1}{5} + \cdots = \frac{2 \cdot 3 \cdot 5 \cdot 7 \cdot 11 \cdot 13 \cdots}{1 \cdot 2 \cdot 4 \cdot 6 \cdot 10 \cdot 12 \cdots},$$

"where," he explained, "the numerator on the right is the product of all the primes and the denominator is the product of all numbers one less than the primes."[4]

Euler began his "proof" by letting $x = 1 + \frac{1}{2} + \frac{1}{3} + \frac{1}{4} + \frac{1}{5} + \cdots$. Although x, being infinite, is not a number at all, he treated it with the usual rules of algebra. Dividing by 2 and subtracting yielded:

$$\frac{1}{2}x = x - \frac{1}{2}x = \left[1 + \frac{1}{2} + \frac{1}{3} + \frac{1}{4} + \frac{1}{5} + \cdots \right] - \left[\frac{1}{2} + \frac{1}{4} + \frac{1}{6} + \frac{1}{8} + \cdots \right]$$

$$= 1 + \frac{1}{3} + \frac{1}{5} + \frac{1}{7} + \frac{1}{9} + \cdots, \tag{4.2}$$

"in which," he observed, "no denominators are even." Dividing series (4.2) by 3, Euler found

$$\frac{1}{3}\left[\frac{1}{2}x \right] = \frac{1}{3}\left[1 + \frac{1}{3} + \frac{1}{5} + \frac{1}{7} + \frac{1}{9} + \cdots \right] = \frac{1}{3} + \frac{1}{9} + \frac{1}{15} + \frac{1}{21} + \frac{1}{27} + \cdots,$$

---

[4] Ibid., pp. 227–229.

and subtracting this from (4.2) produced:

$$\frac{1}{2}x - \frac{1}{3}\left[\frac{1}{2}x\right] = \left[1 + \frac{1}{3} + \frac{1}{5} + \frac{1}{7} + \frac{1}{9} + \cdots\right]$$
$$- \left[\frac{1}{3} + \frac{1}{9} + \frac{1}{15} + \frac{1}{21} + \frac{1}{27} + \cdots\right],$$

or simply

$$\frac{1 \cdot 2}{2 \cdot 3}x = 1 + \frac{1}{5} + \frac{1}{7} + \frac{1}{11} + \frac{1}{13} + \cdots$$

"whose denominators are divisible by neither 2 nor 3." Extending the process to the next stage gave:

$$\frac{1 \cdot 2}{2 \cdot 3}x - \frac{1}{5}\left[\frac{1 \cdot 2}{2 \cdot 3}x\right]$$
$$= \left[1 + \frac{1}{5} + \frac{1}{7} + \frac{1}{11} + \frac{1}{13} + \cdots\right] - \left[\frac{1}{5} + \frac{1}{25} + \frac{1}{35} + \frac{1}{55} + \cdots\right],$$

so that

$$\frac{1 \cdot 2 \cdot 4}{2 \cdot 3 \cdot 5}x = 1 + \frac{1}{7} + \frac{1}{11} + \frac{1}{13} + \frac{1}{17} + \cdots$$

To Euler the pattern was clear. At each step we remove another prime and its multiples from the ranks of the denominators. We thereby generate a reduced series starting with $1 + 1/p$, where p is the next unaddressed prime. For one undaunted by infinite processes, Euler concluded that an *infinitude* of such divisions and subtractions would lead to

$$\frac{1 \cdot 2 \cdot 4 \cdot 6 \cdot 10 \cdot 12 \cdot 16 \cdots}{2 \cdot 3 \cdot 5 \cdot 7 \cdot 11 \cdot 13 \cdot 17 \cdots}x = 1,$$

and a final cross multiplication produced the desired result:

$$1 + \frac{1}{2} + \frac{1}{3} + \frac{1}{4} + \frac{1}{5} + \cdots = x = \frac{2 \cdot 3 \cdot 5 \cdot 7 \cdot 11 \cdot 13 \cdots}{1 \cdot 2 \cdot 4 \cdot 6 \cdot 10 \cdot 12 \cdots}.$$

From a modern perspective this argument, with its repeated operations on divergent series, is as porous as the Swiss cheese of Euler's homeland. Yet it has a certain suggestiveness and an undeniable appeal.

To see why, first suppose we are asked to sum the reciprocals of all positive integers whose only prime factors are 2 and 3. That is, we seek

$$S = 1 + \frac{1}{2} + \frac{1}{3} + \frac{1}{4} + \frac{1}{6} + \frac{1}{8} + \frac{1}{9} + \frac{1}{12} + \frac{1}{16} + \frac{1}{18} + \frac{1}{24} + \frac{1}{27} + \frac{1}{32} + \cdots.$$

Here the denominators are all numbers of the form $2^m 3^n$ so we could just as well write

$$S = \left[1 + \frac{1}{2} + \frac{1}{4} + \frac{1}{8} + \cdots + \frac{1}{2^m} + \cdots \right]$$

$$\times \left[1 + \frac{1}{3} + \frac{1}{9} + \frac{1}{27} + \cdots + \frac{1}{3^n} + \cdots \right]$$

$$= \frac{1}{1 - \frac{1}{2}} \times \frac{1}{1 - \frac{1}{3}} = \frac{2 \cdot 3}{1 \cdot 2}.$$

This strand of reasoning can be extended. For instance, the sum of the reciprocals of all numbers whose prime decompositions contain only 2, 3, and 5 will be:

$$1 + \frac{1}{2} + \frac{1}{3} + \frac{1}{4} + \frac{1}{5} + \frac{1}{6} + \frac{1}{8} + \frac{1}{9} + \frac{1}{10} + \frac{1}{12} + \frac{1}{15}$$

$$+ \frac{1}{16} + \frac{1}{18} + \frac{1}{20} + \frac{1}{24} + \frac{1}{25} + \cdots$$

$$= \frac{1}{1 - \frac{1}{2}} \times \frac{1}{1 - \frac{1}{3}} \times \frac{1}{1 - \frac{1}{5}} = \frac{2 \cdot 3 \cdot 5}{1 \cdot 2 \cdot 4}.$$

And why stop there? After all, *every* whole number is uniquely expressible as the product of primes, and therefore

$$1 + \frac{1}{2} + \frac{1}{3} + \frac{1}{4} + \frac{1}{5} + \frac{1}{6} + \frac{1}{7} + \cdots$$

$$= \left[1 + \frac{1}{2} + \frac{1}{2^2} + \frac{1}{2^3} + \cdots \right] \times \left[1 + \frac{1}{3} + \frac{1}{3^2} + \frac{1}{3^3} + \cdots \right]$$

$$\times \left[1 + \frac{1}{5} + \frac{1}{5^2} + \frac{1}{5^3} + \cdots \right] \times \cdots$$

$$= \frac{1}{1 - \frac{1}{2}} \times \frac{1}{1 - \frac{1}{3}} \times \frac{1}{1 - \frac{1}{5}} \times \frac{1}{1 - \frac{1}{7}} \times \frac{1}{1 - \frac{1}{11}} \times \cdots$$

$$= \frac{2 \cdot 3 \cdot 5 \cdot 7 \cdot 11 \cdot 13 \cdots}{1 \cdot 2 \cdot 4 \cdot 6 \cdot 10 \cdot 12 \cdots},$$

which is precisely the conclusion Euler had reached.

In modern notation, his result would be expressed as

$$\sum_{k=1}^{\infty} \frac{1}{k} = \prod_{p} \frac{1}{1 - \frac{1}{p}},$$

where the (divergent) sum on the left extends over all positive integers and the (divergent) product on the right extends over all prime numbers. Of course, care must be taken to patch up the logical shortcomings. This was provided in 1876 by Leopold Kronecker (1823–1891).[5] He proved that

$$\sum_{k=1}^{\infty} \frac{1}{k^s} = \prod_p \frac{1}{1 - 1/p^s}$$

for $s > 1$ and then interpreted Euler's "theorem" as the outcome of letting $s \to 1^+$.

Putting aside questions of rigor, we see that Euler's powerful intuition had bridged the chasm between the harmonic series and prime numbers—that is, between analysis and number theory. Having crossed this bridge, mathematicians were in no mood to go back.

Consider, for instance, this consequence of Euler's result:

Corollary. *There are infinitely many primes.*

Proof. As we know, $\sum_{k=1}^{\infty} 1/k$ is infinite. So, therefore, is

$$\prod_p \frac{1}{1 - \frac{1}{p}},$$

which could happen only if the number of factors in this product—and hence the number of primes—were infinite.   Q.E.D.

Of course it is not the conclusion here that is new, for it is the same one Euclid had established 2,000 years before. What makes *this* proof so memorable is the means employed to reach its end, namely to deduce the infinitude of primes from the divergence of the harmonic series—a strikingly original idea.

In the same 1737 paper, Euler's attention was directed to a far more subtle theorem about the distribution of primes, one that requires a word or two of introduction. Obviously the sum of all primes

$$2 + 3 + 5 + 7 + 11 + 13 + 17 + \cdots$$

is infinite. Far less apparent is the behavior of the sum of the *reciprocals* of the primes:

$$\frac{1}{2} + \frac{1}{3} + \frac{1}{5} + \frac{1}{7} + \frac{1}{11} + \frac{1}{13} + \frac{1}{17} + \cdots.$$

---

[5]Dickson, p. 413.

On the one hand, this infinite series could behave like the harmonic series and diverge. This would suggest that the primes are reasonably "plentiful" in their distribution among the whole numbers. On the other hand, the series could resemble $\sum_{k=1}^{\infty} 1/k^2$ and converge to a finite sum. This would be the case if the primes—like the squares—were relatively uncommon among the integers. Which situation holds for $\sum_p 1/p$? This is the question that Euler posed.

For notational ease, he let $M = \sum_{k=1}^{\infty} 1/k$ be the harmonic series. By the theorem above, Euler knew that

$$M = \frac{2 \cdot 3 \cdot 5 \cdot 7 \cdot 11 \cdot 13 \cdots}{1 \cdot 2 \cdot 4 \cdot 6 \cdot 10 \cdot 12 \cdots} = \frac{1}{\frac{1}{2} \times \frac{2}{3} \times \frac{4}{5} \times \frac{6}{7} \times \frac{10}{11} \times \cdots}.$$

He then did what came naturally—he took logarithms of both sides to get:

$$\ln M = -\ln(1/2) - \ln(2/3) - \ln(4/5) - \ln(6/7) - \ln(10/11) - \cdots$$

$$= -\ln(1 - 1/2) - \ln(1 - 1/3) - \ln(1 - 1/5)$$

$$- \ln(1 - 1/7) - \ln(1 - 1/11) - \cdots.$$

Each of these Euler expanded using the series

$$\ln(1 - x) = -x - \frac{x^2}{2} - \frac{x^3}{3} - \frac{x^4}{4} - \cdots$$

that we derived in Chapter 2. This gave him an infinitude of infinite series:

$$\ln M = \frac{1}{2} + \frac{1}{2}\left(\frac{1}{2}\right)^2 + \frac{1}{3}\left(\frac{1}{2}\right)^3 + \frac{1}{4}\left(\frac{1}{2}\right)^4 + \frac{1}{5}\left(\frac{1}{2}\right)^5 + \cdots$$

$$+ \frac{1}{3} + \frac{1}{2}\left(\frac{1}{3}\right)^2 + \frac{1}{3}\left(\frac{1}{3}\right)^3 + \frac{1}{4}\left(\frac{1}{3}\right)^4 + \frac{1}{5}\left(\frac{1}{3}\right)^5 + \cdots$$

$$+ \frac{1}{5} + \frac{1}{2}\left(\frac{1}{5}\right)^2 + \frac{1}{3}\left(\frac{1}{5}\right)^3 + \frac{1}{4}\left(\frac{1}{5}\right)^4 + \frac{1}{5}\left(\frac{1}{5}\right)^5 + \cdots$$

$$+ \frac{1}{7} + \frac{1}{2}\left(\frac{1}{7}\right)^2 + \frac{1}{3}\left(\frac{1}{7}\right)^3 + \frac{1}{4}\left(\frac{1}{7}\right)^4 + \frac{1}{5}\left(\frac{1}{7}\right)^5 + \cdots$$

$$\vdots \quad \vdots \quad \quad \vdots \quad \quad \vdots \quad \quad \vdots \quad \quad \vdots$$

Euler summed down the columns:

$$\ln M = \left[\frac{1}{2} + \frac{1}{3} + \frac{1}{5} + \frac{1}{7} + \frac{1}{11} + \frac{1}{13} + \cdots\right]$$

$$+ \frac{1}{2}\left[\left(\frac{1}{2}\right)^2 + \left(\frac{1}{3}\right)^2 + \left(\frac{1}{5}\right)^2 + \left(\frac{1}{7}\right)^2 + \left(\frac{1}{11}\right)^2 + \cdots\right]$$

$$+ \frac{1}{3}\left[\left(\frac{1}{2}\right)^3 + \left(\frac{1}{3}\right)^3 + \left(\frac{1}{5}\right)^3 + \left(\frac{1}{7}\right)^3 + \left(\frac{1}{11}\right)^3 + \cdots\right]$$

$$+ \frac{1}{4}\left[\left(\frac{1}{2}\right)^4 + \left(\frac{1}{3}\right)^4 + \left(\frac{1}{5}\right)^4 + \left(\frac{1}{7}\right)^4 + \left(\frac{1}{11}\right)^4 + \cdots\right]$$

$$+ \frac{1}{5}\left[\left(\frac{1}{2}\right)^5 + \left(\frac{1}{3}\right)^5 + \left(\frac{1}{5}\right)^5 + \left(\frac{1}{7}\right)^5 + \left(\frac{1}{11}\right)^5 + \cdots\right]$$

$$\vdots \quad \vdots \qquad \vdots \qquad \vdots \qquad \vdots \qquad \vdots$$

This he wrote more concisely as $\ln M = A + \frac{1}{2}B + \frac{1}{3}C + \frac{1}{4}D + \frac{1}{5}E + \cdots$, where

$$A = \sum_p \frac{1}{p}, \quad B = \sum_p \frac{1}{p^2}, \quad C = \sum_p \frac{1}{p^3},$$

and so on, with the sums taken over all primes.

At this point, Euler observed, almost casually, "Not only do B, C, D, etc. have finite values, but $\frac{1}{2}B + \frac{1}{3}C + \frac{1}{4}D + \frac{1}{5}E + \cdots$ has a finite value as well." He then moved on to wrap up the proof.[6]

Not so fast, Leonhard! Although these observations may have been evident to *him*, we should insert a brief digression to verify his claim. Fortunately, this can be done with two simple lemmas:

Lemma 1. *For* $n \geq 2$, $\displaystyle\sum_{k=2}^{\infty} \frac{1}{k^n} \leq \frac{1}{n-1}$.

Proof. Considering the shaded rectangles beneath the graph of $y = 1/x^n$ in Figure 4.1, we see that

$$\sum_{k=2}^{\infty} \frac{1}{k^n} = \text{Shaded Area} \leq \int_1^{\infty} \frac{1}{x^n}\,dx = \frac{1}{n-1}.$$

---

[6]Euler, *Opera Omnia*, Ser. 1, Vol. 14, p. 243.

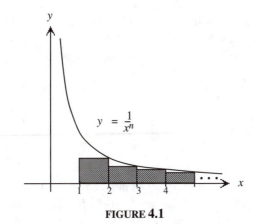

FIGURE 4.1

Note that this verifies Euler's comment that "*B, C, D*, etc. have finite values" because for any $n \geq 2$,

$$\sum_p \frac{1}{p^n} \leq \sum_p \frac{1}{p^2} \leq \sum_{k=2}^{\infty} \frac{1}{k^2} \leq 1 < \infty. \qquad \text{Q.E.D.}$$

Lemma 2. $\frac{1}{2}B + \frac{1}{3}C + \frac{1}{4}D + \frac{1}{5}E + \cdots$ *is finite.*

Proof.

$$\frac{1}{2}B + \frac{1}{3}C + \frac{1}{4}D + \cdots = \frac{1}{2}\sum_p \frac{1}{p^2} + \frac{1}{3}\sum_p \frac{1}{p^3} + \frac{1}{4}\sum_p \frac{1}{p^4} + \cdots$$

$$\leq \frac{1}{2}\sum_{k=2}^{\infty} \frac{1}{k^2} + \frac{1}{3}\sum_{k=2}^{\infty} \frac{1}{k^3} + \frac{1}{4}\sum_{k=2}^{\infty} \frac{1}{k^4} + \cdots$$

$$\leq \frac{1}{2}(1) + \frac{1}{3}\left(\frac{1}{2}\right) + \frac{1}{4}\left(\frac{1}{3}\right) + \frac{1}{5}\left(\frac{1}{4}\right) + \cdots \qquad \text{by Lemma 1}$$

$$\leq 1 + \frac{1}{2}\left(\frac{1}{2}\right) + \frac{1}{3}\left(\frac{1}{3}\right) + \frac{1}{4}\left(\frac{1}{4}\right) + \cdots = \sum_{k=1}^{\infty} \frac{1}{k^2} = \frac{\pi^2}{6} < \infty.$$

So, Euler—although rather unforthcoming on this point in his 1737 paper—was indeed correct. Q.E.D.

We now return to Euler's main result. He stated it as, "The sum of the reciprocals of the prime numbers ... is infinite, an infinity nevertheless smaller than the sum of the harmonic series."[7] In modern terminology, this becomes:

Theorem. $\sum_p 1/p$ *diverges.*

Proof. Because $\ln M = A + \frac{1}{2}B + \frac{1}{3}C + \frac{1}{4}D + \frac{1}{5}E + \cdots$, Euler knew that

$$M = e^{\ln M} = e^{A + \frac{1}{2}B + \frac{1}{3}C + \frac{1}{4}D + \cdots} = e^A \times e^{\frac{1}{2}B + \frac{1}{3}C + \frac{1}{4}D + \cdots}.$$

The term on the left—M—is the harmonic series and thus infinitely large. As a consequence the right-hand side must be infinite too. But Lemma 2 established that $\frac{1}{2}B + \frac{1}{3}C + \frac{1}{4}D + \cdots$ is finite, so $e^{\frac{1}{2}B + \frac{1}{3}C + \frac{1}{4}D + \cdots}$ is finite as well. Because the infinitude of the right side must come from *somewhere*, Euler deduced that e^A is infinite. Hence, $A = \ln(e^A) = \ln(\infty) = \infty$. In his words,

$$\frac{1}{2} + \frac{1}{3} + \frac{1}{5} + \frac{1}{7} + \frac{1}{11} + \frac{1}{13} + \cdots = A = \infty,$$

and the theorem is proved. The series of reciprocals of primes diverges.

Q.E.D.

This proof is a symbol manipulator's dream, an argument that reveals the hand of the master. And it is significant for another reason. In the words of André Weil, "One may well regard these investigations as marking the birth of analytic number theory."[8]

Epilogue

In this epilogue, we have three objectives: to provide an alternate and fully rigorous proof of the divergence of the prime reciprocals; to discuss the infinitude of the $4k + 1$ primes; and to describe briefly the flowering of analytic number theory in the nineteenth century.

Mathematicians who followed Euler—for whom the demands of logical precision were much higher than in his day—often re-proved his theorems according to these more stringent standards. Thus it is not surprising to find alternate proofs of the divergence of $\sum_p 1/p$. In the interest of rigor, we shall present an argument from 1971 due to number theorist Ivan Niven.[9]

---

[7] Ibid., p. 242.
[8] Weil, p. 267.
[9] Ivan Niven, "A Proof of the Divergence of $\sum 1/p$," *The American Mathematical Monthly*, Vol. 78, No. 3, 1971, pp. 272–273.

Before beginning, we observe that any whole number can be written as a product of two factors, one a perfect square and the other "square-free." That is, any n can be uniquely expressed as $n = j^2 k$, where k has no factor (other than 1) that is a perfect square. This observation is self-evident, for upon factoring n into primes, we segregate those that occur in pairs from those that do not. For instance, if $n = 2^5 \cdot 3^4 \cdot 5^2 \cdot 7^3 \cdot 11$, we would decompose

$$n = (2^4 \times 3^4 \times 5^2 \times 7^2) \times (2 \times 7 \times 11) = 1260^2 \times 154,$$

where the second factor is square-free because it contains all different primes.

Following Niven, we adopt a notational convention: let $\sum'_{k \le n} 1/k$ represent the sum of the reciprocals of all *square-free* integers less than or equal to n (including 1). For instance,

$$\sum_{k \le 13}' \frac{1}{k} = 1 + \frac{1}{2} + \frac{1}{3} + \frac{1}{5} + \frac{1}{6} + \frac{1}{7} + \frac{1}{10} + \frac{1}{11} + \frac{1}{13}.$$

With this minimal background, we establish a preliminary lemma and then present Niven's proof.

Lemma. $\displaystyle\lim_{n \to \infty} \left(\sum_{k \le n}' \frac{1}{k} \right) = \infty.$

Proof. We first assert that, for any $n \ge 1$,

$$1 + \frac{1}{2} + \frac{1}{3} + \frac{1}{4} + \frac{1}{5} + \cdots + \frac{1}{n} \le \left(\sum_{j \le n} \frac{1}{j^2} \right) \times \left(\sum_{k \le n}' \frac{1}{k} \right).$$

This follows from the observation above, because any $r \le n$ can be uniquely expressed as $r = j^2 k$, where k is square-free. Therefore $1/r$ appears once and only once in the product on the right. Of course, this product contains more than just $1 + \frac{1}{2} + \frac{1}{3} + \cdots + \frac{1}{n}$. For instance,

$$\left(\sum_{j \le 13} \frac{1}{j^2} \right) \times \left(\sum_{k \le 13}' \frac{1}{k} \right)$$

generates not only the terms $1 + \frac{1}{2} + \frac{1}{3} + \cdots + \frac{1}{13}$ but also fractions like $\frac{1}{40} = \frac{1}{2^2} \times \frac{1}{10}$ and $\frac{1}{150} = \frac{1}{5^2} \times \frac{1}{6}$. For our purposes, however, the inequality is sufficient.

From the assertion, we deduce that

$$1 + \frac{1}{2} + \frac{1}{3} + \frac{1}{4} + \frac{1}{5} + \cdots + \frac{1}{n} \leq \left(\sum_{j \leq n} \frac{1}{j^2} \right) \times \left(\sum_{k \leq n}' \frac{1}{k} \right)$$

$$\leq \left(\sum_{j=1}^{\infty} \frac{1}{j^2} \right) \times \left(\sum_{k \leq n}' \frac{1}{k} \right)$$

$$= \frac{\pi^2}{6} \times \left(\sum_{k \leq n}' \frac{1}{k} \right),$$

where Euler's summation from the previous chapter puts in yet another appearance.

Hence for all n, $\sum_{k \leq n}' 1/k \geq 6/\pi^2 (1 + \frac{1}{2} + \frac{1}{3} + \frac{1}{4} + \cdots + \frac{1}{n})$, and the divergence of the harmonic series guarantees that $\lim_{n \to \infty} (\sum_{k \leq n}' 1/k) = \infty$. In words, the sum of the reciprocals of all square-free integers diverges.

Q.E.D.

Theorem. $\sum_p 1/p$ *diverges.*

Proof (by contradiction). Suppose instead that $\sum_p 1/p = A < \infty$. Recall in Chapter 2 we saw Euler's expansion of $e^x = 1 + x + x^2/2! + x^3/3! + \cdots$. It follows, for $x > 0$, that $e^x \geq 1 + x$. Now let $n \geq 2$ be any whole number and let q be the largest prime less than or equal to n. Then

$$e^A > e^{1/2+1/3+1/5+1/7+\cdots+1/q} = \prod_{p \leq n} e^{1/p} \geq \prod_{p \leq n} \left(1 + \frac{1}{p} \right)$$

$$= \left(1 + \frac{1}{2} \right) \left(1 + \frac{1}{3} \right) \left(1 + \frac{1}{5} \right) \left(1 + \frac{1}{7} \right) \cdots \left(1 + \frac{1}{q} \right) \geq \sum_{k \leq n}' \frac{1}{k}.$$

This last inequality follows because the product under consideration, where no prime is repeated, generates the reciprocals of all square-free integers up to n (along with larger square-free reciprocals as well). But we then must conclude that for any $n \geq 2$,

$$\sum_{k \leq n}' \frac{1}{k} < e^A < \infty,$$

a contradiction because, as the previous lemma established, the series on the left diverges. We are led again to Euler's theorem: the sum of the reciprocals of the primes is itself a divergent series.

Q.E.D.

This modern proof, carefully handling the issue of divergent series in a logically impeccable fashion, provides a counterpoint to Euler's more freewheeling argument. The proofs demonstrate how mathematicians of two different centuries arrived at the same destination. It must be conceded, of course, that *we* have the advantage of traveling in Euler's well-worn tracks; he was blazing a new trail entirely.

As second topic of this epilogue, we return to primes of the form $4k + 1$. A paper of Euler's from 1775 bears on the question of their abundance.[10] There he considered the infinite series

$$\frac{1}{3} - \frac{1}{5} + \frac{1}{7} + \frac{1}{11} - \frac{1}{13} - \frac{1}{17} + \frac{1}{19} + \frac{1}{23} - \frac{1}{29} + \cdots$$

containing reciprocals of the odd primes, with positive signs preceding the $4k - 1$ primes and negative signs preceding the $4k + 1$ primes. After some typically dazzling series manipulations, he approximated the sum by 0.3349816, which, if not terribly accurate, at least convinced him that the series converged to a number somewhere around one-third. With this, he asserted that there were infinitely many of the $4k + 1$ primes. Here is a justification of that claim:

Let

$$S = \frac{1}{5} + \frac{1}{13} + \frac{1}{17} + \frac{1}{29} + \frac{1}{37} + \frac{1}{41} + \cdots$$

and

$$T = \frac{1}{3} + \frac{1}{7} + \frac{1}{11} + \frac{1}{19} + \frac{1}{23} + \frac{1}{31} + \cdots$$

be the series of prime reciprocals grouped by family. Obviously

$$T = S + \left(\frac{1}{3} - \frac{1}{5} + \frac{1}{7} + \frac{1}{11} - \frac{1}{13} - \frac{1}{17} + \frac{1}{19} + \frac{1}{23} - \frac{1}{29} + \cdots \right)$$
$$\approx S + 0.3349816,$$

and so

$$\sum_p \frac{1}{p} = \frac{1}{2} + T + S \approx \frac{1}{2} + 2S + 0.3349816.$$

We have seen (twice) that the left-hand side of this expression is infinite. Therefore, $S = \frac{1}{5} + \frac{1}{13} + \frac{1}{17} + \frac{1}{29} + \frac{1}{37} + \frac{1}{41} + \cdots$ must diverge, a phenomenon that will occur only if there are infinitely many primes of the form $4k + 1$.

---

[10]Euler, *Opera Omnia*, Ser. 1, Vol. 4, pp. 146–162.

Simultaneously, Euler made another, more ambitious conjecture: that if one chooses primes of the form $100k + 1$—the first few of which are 101, 401, 601, 701, 1201—then the sum of their reciprocals is likewise infinite; it would follow that there are infinitely many primes of this type.[11] A natural, and comprehensive, generalization of his ideas is that *any* arithmetic progression

$$a, a + b, a + 2b, a + 3b, \cdots, a + kb, \cdots$$

contains within it infinitely many prime numbers (where we attach the trivial restriction that a and b are relatively prime).

Euler did not prove this conjecture. Indeed, it remained open well into the nineteenth century. The eventual proof in 1837 by Peter Gustav Lejeune-Dirichlet (1805–1859) was like a trumpet blast announcing the arrival of analytic number theory as a mature and powerful subdiscipline.[12]

Among other things, Dirichlet's theorem guaranteed that infinitely many primes occur in the arithmetic progressions

$$1, 5, 9, 13, 17, \ldots, 4k + 1, \ldots \quad \text{and} \quad 3, 7, 11, 15, 19, \ldots, 4k - 1, \ldots,$$

and it therefore *simultaneously* proved the infinitude of both types of primes. This was obviously a formidable result. And, although more far-reaching than Euler's original work, Dirichlet's success owed much to his illustrious predecessor.

As the nineteenth century progressed, analytic number theorists had one objective beyond all others: a proof of the so-called "prime number theorem." Here mathematicians returned to the seemingly intractable mystery of how the primes are distributed among the whole numbers. The prime number theorem identified a pattern—at least approximately. We end this chapter with a look at one of the most profound theorems in all of mathematics.[13]

One way to assess the distribution of primes is to take an inventory. That is, determine the proportion of numbers below 100 that are prime, then do the same for those below 1,000, or 1,000,000. A notational convention is to let $\pi(x)$ be the number of primes less than or equal to x. Then $\pi(x)/x$ is the proportion in question.

A quick check reveals that $\pi(10) = 4$ because 2, 3, 5, and 7 are the primes at or below 10 . Likewise $\pi(100) = 25$, $\pi(1000) = 168$, and $\pi(1,000,000) =$

[11] Ibid., p. 147.

[12] G. Lejeune Dirichlet, *Werke*, Vol. 1, Berlin, 1889, pp. 315–342.

[13] See L. J. Goldstein, "A History of the Prime Number Theorem," *The American Mathematical Monthly*, Vol. 80, No. 6, 1973, pp. 599–614.

$78,498$. This means that 40% of the numbers below 10 are prime; that 25% below 100 are prime; and that 7.85% below 1,000,000 are prime. The relative frequency of primes evidently declines as x grows, but what law describes this trend?

In the last years of the eighteenth century, the young Gauss made a conjecture about the behavior of the proportion $\pi(x)/x$ as x grows without bound. He suggested that $\pi(x)/x \approx 1/\ln x$ when x is large—or, in the language of limits,

$$\lim_{x \to \infty} \frac{\pi(x)}{x/\ln x} = 1.$$

For instance, if $x = 1,000,000$, the exact proportion of primes below x is $\pi(1000000)/1000000 = 0.078498$ whereas $1/\ln(1000000) = 0.072382$. The agreement here is far from perfect, but then 1,000,000 is far from infinity. Agreement improves as x increases.

The transformation of this result from "conjecture" to "theorem" took almost exactly 100 years. The long delay was not for lack of trying. Mathematicians like Legendre (1752–1833), Riemann (1826–1866), and Chebyshev (1821–1894) had first to hone the tools of analytic number theory to a sufficient sharpness for the job at hand. Finally, in 1896, Jacques Hadamard (1865–1963) and C. J. de la Vallée Poussin (1866–1962) simultaneously and independently furnished the proof, and analytic number theory achieved its finest triumph.

Those familiar with the prime number theorem may forget how wondrous a thing it is, linking the primes to the natural logarithm function. Yet this is precisely the sort of connection—between the discrete and the continuous—that Euler first perceived in the propositions examined above.

With the prime number theorem, we conclude our account of these strange mathematical bedfellows. We hope that a case has been made to justify the Shakespearean lines that appeared at the beginning of the chapter. And we hope the reader is ready to acknowledge the debt owed to Leonhard Euler by Dirichlet and Hadamard and Vallée Poussin. If Euler does not quite deserve to be called the "parent" of analytic number theory, let us at least credit him with being its obvious grandparent.

CHAPTER 5

Euler and Complex Variables

In his 1637 masterpiece *Géométrie*, René Descartes addressed the perplexing matter of square roots of negative numbers. "Neither the true (i.e., positive) nor the false (i.e., negative) roots are always real," he wrote, "sometimes they are imaginary."[1]

The term "imaginary" is hardly one to inspire confidence. It sounds slightly delusional, as though a discussion of imaginaries ought to begin with the phrase "Once upon a time." Imaginary numbers seem unlikely to have any real significance.

Nothing could be farther from the truth. Once mathematicians overcame their squeamishness about square roots of negatives, they discovered that such entities played a critical role in mathematics. Complex numbers (as they now prefer to be called) were anything but a worthless sidebar. On the contrary, the complex realm not only offered exciting new challenges but also provided unexpected information about the real numbers embedded within. Mathematicians thereby saw the familiar through an unfamiliar—but indisputably useful—lens.

Not surprisingly, a full understanding and acceptance of complex variables did not happen at once. In this chapter we shall consider the origins of the subject before examining the discoveries of one of its great pioneers, Leonhard Euler, whose mathematical *imagination* was perfectly suited to the mathematical imaginary.

Prologue

It would seem reasonable to assume that square roots of negative numbers first received serious attention when mathematicians grappled with quadratic

---

[1] René Descartes, *The Geometry of René Descartes*, trans. David Eugene Smith and Marcia Latham, Dover, New York, 1954, p. 175.

equations. The solution of $x^2 + 1 = 0$, after all, leads directly to $x^2 = -1$ and then to $x = \pm\sqrt{-1}$.

If one made such an assumption, one would be wrong. In fact, mathematicians readily dismissed equations like $x^2 + 1 = 0$ as being unsolvable if not ridiculous. There was no need to waste time on a problem so absurd, any more than one should try to solve $e^x = -1$ or $\cos x = 2$. Such things were impossible. Case closed.

Rather, imaginary numbers got their foot in the mathematical door because of the problem of *cubic* equations. That is, imaginaries proved unavoidable when dealing with the real solutions of real cubics. This phenomenon, as important as it was unanticipated, is where our story must begin.

Suppose we wish to solve the cubic $x^3 = 6x + 4$ (an example taken from Euler and to which we shall return later in this chapter).[2] For us, a natural first step is to graph the function $f(x) = x^3 - 6x - 4$ and look for x-intercepts, as shown in Figure 5.1.

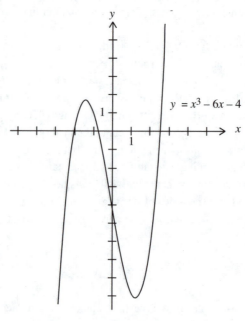

FIGURE 5.1

[2]Euler, *Elements of Algebra*, p. 268.

The graph makes it clear that our cubic has three real solutions—two negative and one positive. Of course, all cubic graphs share the feature of having some points above and some points below the x-axis. Because cubics are continuous, the intermediate value theorem guarantees that at some point the cubic must intersect that axis, and the resulting x-intercept will be a solution to the equation. Consequently, we know that *every* real cubic has at least one real solution.

In sixteenth-century Italy—before the appearance of analytic geometry—mathematicians sought explicit solutions to equations, not vague assertions of "existence." Moreover, their solutions had to be expressed verbally, without benefit of the algebraic symbolism whose appearance still lay decades in the future. The goal, nevertheless, was ambitious: to provide a fail-safe recipe that would yield a solution whenever one existed.

Their model was the quadratic formula, which (in modern notation) says that the equation $ax^2 + bx + c = 0$ has solutions

$$x = \frac{-b \pm \sqrt{b^2 - 4ac}}{2a}.$$

Such a result—employing only the coefficients a, b, and c of the original equation and the basic algebraic operations of addition, subtraction, multiplication, division, and root extraction—is called a "solution by radicals."

The challenge for the Italian algebraists of the 1500s was to find a cubic counterpart to the quadratic formula, a solution by radicals for third-degree equations. In 1515 Scipione del Ferro (1465–1526) of Bologna discovered just that for a special case, the so-called "depressed cubic." This is a third-degree equation lacking a second degree term—that is, one of the form $x^3 = mx + n$.

For the sake of completeness, we shall derive a solution. But rather than consider an archaic sixteenth-century argument, we turn to a clever derivation taken, fittingly enough, from Euler's 1770 textbook, *Elements of Algebra*.[3]

Theorem. *A solution to the depressed cubic $x^3 = mx + n$ is given by*

$$x = \sqrt[3]{\frac{n}{2} + \sqrt{\frac{n^2}{4} - \frac{m^3}{27}}} + \sqrt[3]{\frac{n}{2} - \sqrt{\frac{n^2}{4} - \frac{m^3}{27}}}.$$

Proof. Letting $x = \sqrt[3]{p} + \sqrt[3]{q}$, Euler cubed both sides to get:

---

[3] Ibid., pp. 263–264.

$$x^3 = p + 3\sqrt[3]{p^2 q} + 3\sqrt[3]{pq^2} + q$$
$$= 3\sqrt[3]{pq}\left(\sqrt[3]{p} + \sqrt[3]{q}\right) + (p + q) = \left(3\sqrt[3]{pq}\right)x + (p + q).$$

The resulting equation—i.e., $x^3 = (3\sqrt[3]{pq})x + (p + q)$—has the identical structure of the original depressed cubic $x^3 = mx + n$. This suggests an algebraic strategy of setting $3\sqrt[3]{pq} = m$ and $p + q = n$; from this determining p and q in terms of m and n; and then presenting the solution by radicals as $x = \sqrt[3]{p} + \sqrt[3]{q}$.

Euler did just that. From $3\sqrt[3]{pq} = m$, it follows that $4pq = 4m^3/27$, and from $p + q = n$, it follows that $p^2 + 2pq + q^2 = n^2$. Combining these results yields

$$(p^2 + 2pq + q^2) - 4pq = n^2 - \frac{4m^3}{27}, \quad \text{or simply} \quad (p - q)^2 = n^2 - \frac{4m^3}{27}.$$

Thus $p - q = \sqrt{n^2 - (4m^3/27)}$. To this Euler both added and subtracted $p + q = n$ to get

$$2p = n + \sqrt{n^2 - \frac{4m^3}{27}} \quad \text{and} \quad 2q = n - \sqrt{n^2 - \frac{4m^3}{27}},$$

and so the solution of the original cubic is

$$x = \sqrt[3]{p} + \sqrt[3]{q} = \sqrt[3]{\frac{n}{2} + \sqrt{\frac{n^2}{4} - \frac{m^3}{27}}} + \sqrt[3]{\frac{n}{2} - \sqrt{\frac{n^2}{4} - \frac{m^3}{27}}}. \qquad \text{Q.E.D.}$$

"To render this more clear," (as Euler was fond of saying), we solve the equation $x^3 = 6x + 9$. Here $m = 6$ and $n = 9$, so that

$$x = \sqrt[3]{\frac{9}{2} + \sqrt{\frac{81}{4} - \frac{216}{27}}} + \sqrt[3]{\frac{9}{2} - \sqrt{\frac{81}{4} - \frac{216}{27}}}$$

$$= \sqrt[3]{\frac{9}{2} + \sqrt{\frac{49}{4}}} + \sqrt[3]{\frac{9}{2} - \sqrt{\frac{49}{4}}}$$

$$= \sqrt[3]{8} + \sqrt[3]{1} = 2 + 1 = 3,$$

which certainly checks.

The fact that this formula applies only to *depressed* cubics is not a serious restriction. In the middle of the sixteenth century, it was shown by Girolamo Cardano (1501–1576) that a general cubic equation $z^3 + az^2 + bz + c = 0$ could be transformed into a related, depressed cubic $x^3 = mx + n$ by the substitution $z = x - a/3$. A solution for x in the depressed cubic (by the procedure described above) provided an immediate solution for z in the original. Thus, solving depressed cubics turned out to be sufficient.

After a series of episodes too complicated and bizarre to recount here, Cardano was the first to publish these discoveries in his 1545 work *Ars Magna*. The procedure for solving cubics has since become known as "Cardano's formula."[4]

Thus far, everything seems in order.

But consider our earlier example $x^3 = 6x + 4$, which differs only slightly from the cubic $x^3 = 6x + 9$ that we just solved. Applying the formula with $m = 6$ and $n = 4$, we get

$$x = \sqrt[3]{\frac{4}{2} + \sqrt{\frac{16}{4} - \frac{216}{27}}} + \sqrt[3]{\frac{4}{2} - \sqrt{\frac{16}{4} - \frac{216}{27}}}$$

$$= \sqrt[3]{2 + \sqrt{-4}} + \sqrt[3]{2 - \sqrt{-4}} = \sqrt[3]{2 + 2\sqrt{-1}} + \sqrt[3]{2 - 2\sqrt{-1}}.$$

This answer, containing the square root of a negative number, is perplexing. It suggests on the face of it that a solution does not exist. Yet our graph showed that the cubic has not one but three real solutions. Somewhere, something appears to have gone awry.

Mathematicians were faced with one of two possibilities: either Cardano's formula was incorrect, unreliable, and of limited value; or this imaginary answer was in fact a real number traveling *incognito*.

In his 1570 *Algebra*, Rafael Bombelli (ca. 1526–1573) endorsed the second option. He advanced the idea that we can transform these complex numbers into a real solution of the depressed cubic. Imaginaries would thus be a temporary annoyance whose implausibility could be overlooked when they salvaged Cardano's formula. It was a bold, if not altogether well-understood, idea.

For instance, suppose we suppress any misgivings about square roots of negatives and expand algebraically:

$$(-1 + \sqrt{-1})^3 = (-1)^3 + 3(-1)^2\sqrt{-1} + 3(-1)(\sqrt{-1})^2 + (\sqrt{-1})^3$$

$$= -1 + 3\sqrt{-1} + 3 - \sqrt{-1} = 2 + 2\sqrt{-1}.$$

---

[4] See Girolamo Cardano, *Ars Magna*, trans. T. Richard Witmer, Dover, New York, 1968, pp. 96–101 or Dunham, *Journey Through Genius*, Ch. 6.

From this it would be reasonable to conclude that $\sqrt[3]{2 + 2\sqrt{-1}} = -1 + \sqrt{-1}$. Similarly, we see that $(-1 - \sqrt{-1})^3 = 2 - 2\sqrt{-1}$ and deduce that $\sqrt[3]{2 - 2\sqrt{-1}} = -1 - \sqrt{-1}$.

This sheds new light on the output of Cardano's formula in the example above, for we now have

$$x = \sqrt[3]{2 + 2\sqrt{-1}} + \sqrt[3]{2 - 2\sqrt{-1}} = (-1 + \sqrt{-1}) + (-1 - \sqrt{-1}) = -2.$$

Thus x is a real number after all, and one that is clearly a solution to $x^3 = 6x + 4$. Cardano's formula has been saved.

Or has it? At least two questions still beg for answers. First, how do we know at the outset which complex number to cube in order to get $2 + 2\sqrt{-1}$? Is there an algorithm to suggest that we try $-1 + \sqrt{-1}$, or must we rely on an inexplicable burst of insight?

Second, where are the *other* real roots? After all, we know this cubic has a trio of solutions. Might the others also be lurking somewhere beneath the surface of Cardano's formula?

Bombelli did not have the answers. "The whole matter," he wrote in a revealing passage, "seems to rest on sophistry rather than on truth."[5] His confusion was understandable, and it was echoed over the next century and a half by mathematicians who, when they used imaginaries, seemed slightly embarrassed about it. Even Leibniz, creator of the calculus, called $\sqrt{-1}$ "that amphibian between being and non-being."[6] It seemed that imaginary numbers just could not get any respect.

In order to make progress, it would take an individual capable of overcoming such prejudices. It required someone with an exceptional agility of mind who combined an unwavering faith in the power of symbols with a tendency toward mathematical boldness. Fortunately, such a person was waiting in the wings.

Enter Euler

In his *Elements of Algebra*, Euler introduced $\sqrt{-1}$ as "... neither nothing, nor greater than nothing, nor less than nothing..." and observed

---

[5] Katz, p. 336.
[6] Kline, p. 254.

... we are led to the idea of numbers, which from their nature are impossible; and therefore they are usually called imaginary quantities,
because they exist merely in the imagination.[7]

Lest anyone take this as a condemnation, he continued:

... notwithstanding this, these numbers present themselves to the
mind; they exist in our imagination and we still have a sufficient idea
of them; ... nothing prevents us from making use of these imaginary
numbers, and employing them in calculation.

Compared to the disparaging words of Bombelli, Descartes, and Leibniz
cited above, this sounds like an unqualified endorsement. And Euler certainly
fulfilled his promise to use imaginaries freely in his research.

For instance, in a 1751 paper he explored what we today call the "roots
of unity."[8] The number 1 (i.e., "unity") has the two square roots ± 1, found by
solving $x^2 - 1 = 0$. Similarly, Euler observed that unity has three cube roots,
arising as the solutions of

$$0 = x^3 - 1 = (x - 1)(x^2 + x + 1).$$

The first factor gives $x = 1$. The second, via an application of the quadratic
formula, yields $x = (-1 \pm \sqrt{-3})/2$. Should anyone doubt that these are cube
roots of 1, simply expand and confirm that $\left[(-1 \pm \sqrt{-3})/2\right]^3 = 1$.

Why stop there? By considering $x^4 - 1 = 0$, Euler showed that the
four fourth-roots of unity are $1, -1, \sqrt{-1}$, and $-\sqrt{-1}$. Somewhat more
complicated was for him to identify the five fifth-roots of unity—i.e., the
solutions of $x^5 - 1 = 0$—as:

$$1, \quad \frac{-1 - \sqrt{5} + \sqrt{-10 + 2\sqrt{5}}}{4}, \quad \frac{-1 - \sqrt{5} - \sqrt{-10 + 2\sqrt{5}}}{4},$$

$$\frac{-1 + \sqrt{5} + \sqrt{-10 - 2\sqrt{5}}}{4}, \quad \text{and} \quad \frac{-1 + \sqrt{5} - \sqrt{-10 - 2\sqrt{5}}}{4}.$$

The last four of these are imaginary. It is worth noting that Euler seemed neither
apologetic nor beset by doubts about the validity of these answers. For him,
imaginaries had become full partners in the mathematical enterprise.

---

[7] Euler, *Elements of Algebra*, p. 43.
[8] Euler, *Opera Omnia*, Ser. 1, Vol. 6, pp. 66–77.

As we shall see, Euler found a beautiful short cut for finding roots of unity, or indeed roots of any number, real or complex. With it, he was able to show that

> any quantity has two square roots, three cube roots, four fourth roots, and so on.[9]

This phenomenon, which surely appealed to Euler's sense of order, required for its justification a result now known as "De Moivre's theorem," which we shall examine shortly. If one looks at the work of Abraham De Moivre (1667–1754), one can find an early version of the theorem.[10] Nonetheless, it is fair to say that it was Euler who first grasped its importance and exploited it in a modern fashion. This theorem of De Moivre/Euler is today a cornerstone of complex algebra.

From this point onward, we shall adopt a notation that Euler standardized—namely, $i = \sqrt{-1}$ as the imaginary unit. This has, of course, become one of the best-known symbols of mathematics.

In his *Introductio*, Euler considered expressions of the form $\cos\theta \pm i\sin\theta$, which appeared in the factorization

$$1 = \cos^2\theta + \sin^2\theta = (\cos\theta + i\sin\theta)(\cos\theta - i\sin\theta).$$

These exhibited a noteworthy multiplicative stability because

$$(\cos\theta \pm i\sin\theta)(\cos\phi \pm i\sin\phi) = (\cos\theta\cos\phi - \sin\theta\sin\phi)$$
$$\pm\, i(\sin\theta\cos\phi + \cos\theta\sin\phi)$$
$$= \cos(\theta + \phi) \pm i\sin(\theta + \phi),$$

by the well-known trig identities. Consequently, for $\theta = \phi$, we have

$$(\cos\theta + i\sin\theta)^2 = \cos(2\theta) + i\sin(2\theta)$$

and

$$(\cos\theta - i\sin\theta)^2 = \cos(2\theta) - i\sin(2\theta).$$

A little thought convinced Euler that the same argument transferred to higher powers. He stated the general result—i.e., De Moivre's theorem—as:

$$(\cos\theta \pm i\sin\theta)^n = \cos(n\theta) \pm i\sin(n\theta) \quad \text{for all} \quad n \geq 1.$$

---

[9] Ibid., p. 118.

[10] David Eugene Smith, *A Source Book in Mathematics*, Dover, New York, 1959, pp. 440–450.

As we shall see, Euler used this result with great success in a number of problems, from finding roots of complex numbers, to developing power series expansions for $\cos x$ and $\sin x$, to deducing one of the most extraordinary identities in all of mathematics. In the process, he made imaginary numbers seem ever more "natural," ever more respectable.

Euler's prescription for finding roots of complex numbers appeared in a long and magnificent 1749 paper titled "*Recherches sur les racines imaginaires des équations*" that serves, even to the present day, as an excellent introduction to complex variables.[11] He began with a non-zero complex number $z = a + bi$ whose nth root is desired. He introduced $c = \sqrt{a^2 + b^2}$ and found an angle θ between $-\pi/2$ and $\pi/2$ such that $\sin \theta = b/c$. Then $\cos \theta = \sqrt{1 - \sin^2 \theta} = \sqrt{(c^2 - b^2)/c^2} = a/c$. It follows that $z = a + bi = c(\cos \theta + i \sin \theta)$.

Euler next asserted that z has n different nth roots, given by

$$\sqrt[n]{c}\left(\cos \frac{\theta + 2\pi k}{n} + i \sin \frac{\theta + 2\pi k}{n}\right) \quad \text{for } k = 0, 1, 2, \ldots, n-1$$

or, equivalently, by

$$\sqrt[n]{c}\left(\cos \frac{\theta - 2\pi k}{n} + i \sin \frac{\theta - 2\pi k}{n}\right) \quad \text{for } k = 0, 1, 2, \ldots, n-1$$

Verification was as simple as raising these expressions to the nth power via De Moivre's theorem:

$$\left[\sqrt[n]{c}\left(\cos \frac{\theta \pm 2\pi k}{n} + i \sin \frac{\theta \pm 2\pi k}{n}\right)\right]^n$$
$$= \left(\sqrt[n]{c}\right)^n\left[\cos\left(n \cdot \frac{\theta \pm 2\pi k}{n}\right) + i \sin\left(n \cdot \frac{\theta \pm 2\pi k}{n}\right)\right]$$
$$= c\left[\cos(\theta \pm 2\pi k) + i \sin(\theta \pm 2\pi k)\right] = c(\cos \theta + i \sin \theta) = z.$$

"It is evident," Euler remarked, "that $\sqrt[n]{a + bi} \ldots$ takes the form $M + Ni$." In other words, roots of complex numbers are again complex numbers. This means—in modern parlance—that the imaginary domain is *closed* under the extraction of roots, a property not shared by the integers, the rationals, or the reals. It ranks among the most algebraically significant properties of the complex numbers.

Note that finding roots of unity now becomes a simple task. We begin with $z = 1 = 1 + 0i$, from which it follows that $c = 1$ and $\theta = 0$. Then the nth

---

[11] Euler, *Opera Omnia*, Ser. 1, Vol. 6, pp. 116–118.

roots of unity—usually written as $\omega_0, \omega_1, \ldots, \omega_{n-1}$—are given by

$$\omega_k = \cos \frac{2\pi k}{n} + i \sin \frac{2\pi k}{n} \quad \text{for } k = 0, 1, 2, \ldots, n - 1.$$

Understanding complex roots also allows us to resolve the remaining quandary associated with Cardano's formula. As we saw above, the formula applied to $x^3 = 6x + 4$ yielded solution

$$x = \sqrt[3]{2 + 2\sqrt{-1}} + \sqrt[3]{2 - 2\sqrt{-1}}.$$

The cube roots appearing here can now be determined. Consider first $\sqrt[3]{2 + 2\sqrt{-1}} = \sqrt[3]{2 + 2i}$, where $a = b = 2$, $c = \sqrt{8}$, and $\theta = \sin^{-1}(b/c) = \sin^{-1}(1/\sqrt{2}) = \pi/4$. The three cube roots of $2 + 2i$—which arise from using angles $\pi/4$, $\pi/4 + 2\pi = 9\pi/4$, and $\pi/4 + 4\pi = 17\pi/4$—are thus

$$\sqrt[3]{\sqrt{8}} \left[\cos \frac{\pi}{12} + i \sin \frac{\pi}{12} \right], \quad \sqrt[3]{\sqrt{8}} \left[\cos \frac{3\pi}{4} + i \sin \frac{3\pi}{4} \right],$$

and

$$\sqrt[3]{\sqrt{8}} \left[\cos \frac{17\pi}{12} + i \sin \frac{17\pi}{12} \right].$$

The roots of $2 - 2\sqrt{-1} = 2 - 2i$ are treated similarly. Here $c = \sqrt{8}$, $\theta = \sin^{-1}(-1/\sqrt{2}) = -\pi/4$, and for the three crucial angles we use $-\pi/4$, $-\pi/4 - 2\pi = -9\pi/4$, and $-\pi/4 - 4\pi = -17\pi/4$. So the cube roots of $2 - 2i$ are

$$\sqrt[3]{\sqrt{8}} \left[\cos \left(-\frac{\pi}{12} \right) + i \sin \left(-\frac{\pi}{12} \right) \right],$$

$$\sqrt[3]{\sqrt{8}} \left[\cos \left(-\frac{3\pi}{4} \right) + i \sin \left(-\frac{3\pi}{4} \right) \right],$$

and

$$\sqrt[3]{\sqrt{8}} \left[\cos \left(-\frac{17\pi}{12} \right) + i \sin \left(-\frac{17\pi}{12} \right) \right].$$

Along with the observation that $\sqrt[3]{\sqrt{8}} = \sqrt{2}$, a few trigonometric identities permit a major simplification. For one thing, $\cos(-\theta) = \cos \theta$, and $\sin(-\theta) = -\sin \theta$. Further, from the double-angle formula—$\cos(2\theta) = 2 \cos^2 \theta - 1$—we conclude that $\cos(\pi/6) = 2\cos^2(\pi/12) - 1$ or

$$\cos\frac{\pi}{12} = \sqrt{\frac{1}{2} + \frac{1}{2}\cos\frac{\pi}{6}} = \sqrt{\frac{1}{2} + \frac{\sqrt{3}}{4}}.$$

Likewise, $\cos(17\pi/12) = -\sqrt{1/2 - \sqrt{3}/4}$. Applying these results to the Cardano solution, we see the three real roots jump out:

$$r_1 = \sqrt[3]{2 + 2\sqrt{-1}} + \sqrt[3]{2 - 2\sqrt{-1}}$$

$$= \sqrt{2}\left[\cos\frac{\pi}{12} + i\sin\frac{\pi}{12}\right] + \sqrt{2}\left[\cos\left(-\frac{\pi}{12}\right) + i\sin\left(-\frac{\pi}{12}\right)\right]$$

$$= \sqrt{2}\left[\cos\frac{\pi}{12} + i\sin\frac{\pi}{12} + \cos\frac{\pi}{12} - i\sin\frac{\pi}{12}\right]$$

$$= \sqrt{2}\left[2\cos\frac{\pi}{12}\right] = 2\sqrt{2}\sqrt{\frac{1}{2} + \frac{\sqrt{3}}{4}} = \sqrt{4 + 2\sqrt{3}}.$$

Next comes:

$$r_2 = \sqrt[3]{2 + 2\sqrt{-1}} + \sqrt[3]{2 - 2\sqrt{-1}}$$

$$= \sqrt{2}\left[\cos\frac{3\pi}{4} + i\sin\frac{3\pi}{4}\right] + \sqrt{2}\left[\cos\left(-\frac{3\pi}{4}\right) + i\sin\left(-\frac{3\pi}{4}\right)\right]$$

$$= \sqrt{2}\left[2\cos\frac{3\pi}{4}\right] = -2 \quad \text{(the one we found above)}.$$

And finally, there is:

$$r_3 = \sqrt[3]{2 + 2\sqrt{-1}} + \sqrt[3]{2 - 2\sqrt{-1}}$$

$$= \sqrt{2}\left[\cos\frac{17\pi}{12} + i\sin\frac{17\pi}{12}\right] + \sqrt{2}\left[\cos\left(\frac{17\pi}{12}\right) + i\sin\left(\frac{17\pi}{12}\right)\right]$$

$$= \sqrt{2}\left[2\cos\frac{17\pi}{12}\right] = -\sqrt{4 - 2\sqrt{3}}.$$

This chain of reasoning required some beautiful teamwork: from Cardano to De Moivre to Euler; from algebra to trigonometry to complex variables. By veering off into imaginary numbers, Cardano's formula generated the three *real* roots of $x^3 = 6x + 4$, namely $\sqrt{4 + 2\sqrt{3}}$, -2, and $-\sqrt{4 - 2\sqrt{3}}$, which of course correspond to the three x-intercepts evident in Figure 5.1. The successful resolution of this problem reinforces Euler's enthusiasm for "employing [imaginaries] in calculation."

It is important to repeat that, in finding the real roots of this cubic, we detoured into the complex numbers. One is reminded of Hadamard's astute observation that "The shortest path between two truths in the real domain passes through the complex domain."[12]

In the *Introductio* Euler used De Moivre's theorem in a very different manner to derive two famous series expansions:[13]

Theorem.

$$\cos x = 1 - \frac{x^2}{1 \cdot 2} + \frac{x^4}{1 \cdot 2 \cdot 3 \cdot 4} - \frac{x^6}{1 \cdot 2 \cdot 3 \cdot 4 \cdot 5 \cdot 6} + \cdots \quad and$$

$$\sin x = x - \frac{x^3}{1 \cdot 2 \cdot 3} + \frac{x^5}{1 \cdot 2 \cdot 3 \cdot 4 \cdot 5} - \cdots.$$

Proof. For any $n \geq 1$, Euler knew that

$$\cos n\theta + i \sin n\theta = (\cos \theta + i \sin \theta)^n \text{ and } \cos n\theta - i \sin n\theta = (\cos \theta - i \sin \theta)^n.$$
$$(5.1)$$

Adding and dividing by 2, he concluded

$$\cos n\theta = \frac{(\cos \theta + i \sin \theta)^n + (\cos \theta - i \sin \theta)^n}{2}.$$

He then expanded the powers on the right by means of the binomial theorem to get:

$$\cos n\theta = \frac{1}{2} \left[\cos^n \theta + \frac{ni \cos^{n-1} \theta \sin \theta}{1} - \frac{n(n-1) \cos^{n-2} \theta \sin^2 \theta}{1 \cdot 2} \right.$$

$$\left. - \frac{n(n-1)(n-2)i \cos^{n-3} \theta \sin^3 \theta}{1 \cdot 2 \cdot 3} + \cdots \right]$$

$$+ \frac{1}{2} \left[\cos^n \theta - \frac{ni \cos^{n-1} \theta \sin \theta}{1} - \frac{n(n-1) \cos^{n-2} \theta \sin^2 \theta}{1 \cdot 2} \right.$$

$$\left. + \frac{n(n-1)(n-2)i \cos^{n-3} \theta \sin^3 \theta}{1 \cdot 2 \cdot 3} + \cdots \right]$$

$$= \cos^n \theta - \frac{n(n-1) \cos^{n-2} \theta \sin^2 \theta}{1 \cdot 2}$$

$$+ \frac{n(n-1)(n-2)(n-3) \cos^{n-4} \theta \sin^4 \theta}{1 \cdot 2 \cdot 3 \cdot 4} - \cdots.$$

---

[12] Kline, p. 626.
[13] Euler, *Introduction to Analysis of the Infinite*, Book I, pp. 106–107.

At this point, Euler "went infinite." That is, he let $x = n\theta$, where n is infinitely *large* and thus $\theta = x/n$ is infinitely *small*. In so doing, he noted that $\cos \theta = 1$ and $\sin \theta = \theta = x/n$—a recognition, in modern terminology, that

$$\lim_{\theta \to 0} \cos \theta = 1 \text{ and } \lim_{\theta \to 0} \frac{\sin \theta}{\theta} = 1.$$

Because n was infinitely large, there surely could be no difference between it and $n - 1$, $n - 2$, $n - 3$, and so on; therefore Euler simply replaced each of these by n.

Such mathematical gyrations seem, to modern tastes, unorthodox. However, they permitted Euler to transform the series above into

$$\cos x = 1^n - \frac{n \cdot n \cdot (1)^{n-2}(x/n)^2}{1 \cdot 2} + \frac{n \cdot n \cdot n \cdot n \cdot (1)^{n-4}(x/n)^4}{1 \cdot 2 \cdot 3 \cdot 4} - \cdots$$

$$= 1 - \frac{x^2}{1 \cdot 2} + \frac{x^4}{1 \cdot 2 \cdot 3 \cdot 4} - \frac{x^6}{1 \cdot 2 \cdot 3 \cdot 4 \cdot 5 \cdot 6} + \cdots,$$

and he thereby "proved" the famous expansion of $\cos x$.

In a similar way—after first *subtracting* the expressions in (5.1) and dividing by 2—Euler concluded that

$$\sin n\theta = \frac{(\cos \theta + i \sin \theta)^n - (\cos \theta - i \sin \theta)^n}{2i},$$

from which he derived the related series

$$\sin x = x - \frac{x^3}{1 \cdot 2 \cdot 3} + \frac{x^5}{1 \cdot 2 \cdot 3 \cdot 4 \cdot 5} - \frac{x^7}{1 \cdot 2 \cdot 3 \cdot 4 \cdot 5 \cdot 6 \cdot 7} + \cdots. \quad \text{Q.E.D.}$$

Euler applied De Moivre's theorem in even more spectacular fashion to prove an extraordinary relationship that bears his name: Euler's identity.[14]

Theorem. *For any real x, $e^{ix} = \cos x + i \sin x$.*

Proof. As above, Euler began with

$$\cos n\theta = \frac{(\cos \theta + i \sin \theta)^n + (\cos \theta - i \sin \theta)^n}{2}.$$

Again, he let n be "an infinite number," so that $\theta = x/n$ is infinitely small and thus $\cos(\theta) = 1$ and $\sin(\theta) = \theta = x/n$. Wholesale substitution produced:

---

[14]Ibid., pp. 111-112.

$$\cos x = \cos n\theta = \frac{(\cos \theta + i \sin \theta)^n + (\cos \theta - i \sin \theta)^n}{2}$$

$$= \frac{(1 + ix/n)^n + (1 - ix/n)^n}{2}. \tag{5.2}$$

As we saw in Chapter 2, Euler equated e^ω and $1 + \omega$ for ω infinitely small. Therefore, if a is a finite number and n is infinitely large, we have

$$e^a = (e^{a/n})^n = \left(1 + \frac{a}{n}\right)^n.$$

Replacing a by the finite (albeit imaginary) quantities ix and $-ix$, he transformed equation (5.2) into

$$\cos x = \frac{e^{ix} + e^{-ix}}{2}.$$

Next, Euler mimicked this procedure for the sine function and derived:

$$\sin x = \sin n\theta = \frac{(\cos \theta + i \sin \theta)^n - (\cos \theta - i \sin \theta)^n}{2i}$$

$$= \frac{(1 + ix/n)^n - (1 - ix/n)^n}{2i} = \frac{e^{ix} - e^{-ix}}{2i}.$$

Finally, adding these results produced the formula that will forever rank among Euler's greatest discoveries:

$$\cos x + i \sin x = \frac{e^{ix} + e^{-ix}}{2} + i\frac{e^{ix} - e^{-ix}}{2i} = e^{ix}. \qquad \text{Q.E.D.}$$

"From these equations," Euler noted with evident satisfaction, "we understand how complex exponentials can be expressed by real sines and cosines." His enthusiasm has been echoed by mathematicians ever since. Few would argue that Euler's identity is among the most beautiful formulas of all.

As was his custom, Euler provided alternative proofs of so important a theorem. He did, after all, seem to operate under the principle that any result worth proving is worth proving *again*. In this spirit, we shall consider two other proofs he gave over the course of his career.

The first, quite bold, requires an acquaintance with integral calculus and an Eulerian confidence in the power of symbols.[15]

Theorem. *For any real x, $e^{ix} = \cos x + i \sin x$.*

---

[15] Euler, *Opera Omnia*, Ser. 1, Vol. 19, pp. 431–432.

Proof. Euler introduced $\sin x = y$ so that $x = \sin^{-1} y = \int dy/\sqrt{1 - y^2}$, the last equality being a familiar antiderivative. Without batting an eye, he made a complex change of variable by letting $y = iz$ and $dy = i\,dz$ to get

$$x = \int \frac{i\,dz}{\sqrt{1 - (iz)^2}} = i \int \frac{dz}{\sqrt{1 + z^2}} = i \ln\left(\sqrt{1 + z^2} + z\right). \qquad (5.3)$$

(One may find this antiderivative by means of trigonometric substitution, or look it up in an integral table, or simply check it by differentiation.) Because $z = y/i = \sin x/i$, it followed that $z^2 = \sin^2 x/i^2 = -\sin^2 x$, and so (5.3) became

$$x = i \ln\left(\sqrt{1 - \sin^2 x} + \frac{\sin x}{i}\right) = i \ln(\cos x - i \sin x).$$

Therefore

$$ix = i^2 \ln(\cos x - i \sin x) = \ln \frac{1}{\cos x - i \sin x} = \ln(\cos x + i \sin x),$$

and from here a simple exponentiation completed his proof:

$$e^{ix} = e^{\ln(\cos x + i \sin x)} = \cos x + i \sin x. \qquad \text{Q.E.D.}$$

Again one sees Euler applying to imaginary quantities the familiar rules for real ones. In the eighteenth century, such an application as much a matter of faith as of logic, but Euler, the great symbol manipulator, was never more in his element.

He gave yet another proof of the identity, one that appeared in an important paper on the logarithms of imaginary numbers that we shall consider in the epilogue.[16] This time Euler stated the theorem as:

$$\cos x + i \sin x = \left(1 + \frac{ix}{n}\right)^n, \text{ where } n \text{ is an infinite number.}$$

As we have seen, the expression on the right was, in his mind, identical to e^{ix}.

Theorem. *For any real* x, $e^{ix} = \cos x + i \sin x$.

Proof. Acknowledging that "... the truth of [the identity] is sufficiently proved elsewhere," Euler used the expansions of $\cos x$ and $\sin x$ developed above, as well as

$$e^x = 1 + x + \frac{x^2}{1 \cdot 2} + \frac{x^3}{1 \cdot 2 \cdot 3} + \frac{x^4}{1 \cdot 2 \cdot 3 \cdot 4} + \cdots,$$

---

[16]Euler, *Opera Omnia*, Ser. 1, Vol. 17, p. 219.

that we saw in Chapter 2. Replacing x by ix in this last series, Euler simply followed the formulas to their natural end:

$$e^{ix} = 1 + ix + \frac{(ix)^2}{1 \cdot 2} + \frac{(ix)^3}{1 \cdot 2 \cdot 3} + \frac{(ix)^4}{1 \cdot 2 \cdot 3 \cdot 4} + \frac{(ix)^5}{1 \cdot 2 \cdot 3 \cdot 4 \cdot 5} + \cdots$$

$$= 1 + ix - \frac{x^2}{1 \cdot 2} - \frac{ix^3}{1 \cdot 2 \cdot 3} + \frac{x^4}{1 \cdot 2 \cdot 3 \cdot 4} + \frac{ix^5}{1 \cdot 2 \cdot 3 \cdot 4 \cdot 5} - \cdots$$

$$= \left[1 - \frac{x^2}{1 \cdot 2} + \frac{x^4}{1 \cdot 2 \cdot 3 \cdot 4} - \cdots \right]$$

$$+ i \left[x - \frac{x^3}{1 \cdot 2 \cdot 3} + \frac{x^5}{1 \cdot 2 \cdot 3 \cdot 4 \cdot 5} - \cdots \right]$$

$$= \cos x + i \sin x. \qquad\qquad \text{Q.E.D.}$$

Voila! Three proofs of Euler's identity should be enough for any skeptic.

We conclude this section with a worthy corollary. If we let $x = \pi$ in this formula, we have

$$e^{i\pi} = \cos \pi + i \sin \pi = -1 + i \cdot 0 = -1. \qquad (5.4)$$

The attentive reader may recall that, near the beginning of this chapter, we cited the example of $e^x = -1$ as an absurd, unsolvable equation. Now, thanks to Euler, we have a solution: $x = i\pi$. Freed from the limitations of the real numbers, he solved the unsolvable.

Of course, (5.4) can be rewritten as $e^{i\pi} + 1 = 0$. As math professors are fond of observing, this equation assembles the five most important constants in mathematics, namely

0—the additive identity
1—the multiplicative identity
π—the circular constant
e—the base of the natural logarithms
i—the imaginary unit

That these five superstar numbers should be related in so simple a manner is truly astonishing. That Euler *recognized* such a relationship is a tribute to his mathematical power.

This curious formula may be a fitting place to take a breath. By now, we should have gained some appreciation for Euler and the complex numbers. From his clever exploitation of De Moivre's formula to his multiple derivations of the identity that carries his name, Euler had legitimized imaginary quantities in a way that was both unprecedented and irreversible.

Epilogue

As has been emphasized so often in this book, Leonhard Euler was a pioneer. Having grown comfortable with the fundamentals of complex numbers, he moved ever deeper into uncharted terrain. For instance, he asked whether one could define the sine and cosine of what he called "an imaginary arc"—a bold thought indeed.[17] And he resolved the long-debated problem of logarithms of complex quantities. In this epilogue, we briefly consider these two innovations.

First, what does one make of $\cos(a + bi)$? Euler approached it in stages, initially considering simply $\cos(bi)$. Applying the series expansion for cosine, he wrote:

$$\cos(bi) = 1 - \frac{(bi)^2}{1 \cdot 2} + \frac{(bi)^4}{1 \cdot 2 \cdot 3 \cdot 4} - \frac{(bi)^6}{1 \cdot 2 \cdot 3 \cdot 4 \cdot 5 \cdot 6} + \cdots$$

$$= 1 + \frac{b^2}{1 \cdot 2} + \frac{b^4}{1 \cdot 2 \cdot 3 \cdot 4} + \frac{b^6}{1 \cdot 2 \cdot 3 \cdot 4 \cdot 5 \cdot 6} + \cdots = \frac{e^b + e^{-b}}{2}.$$

This last equality, which Euler spotted at once, can be verified by replacing both exponentials with their series expansions and simplifying. It follows, quite unexpectedly, that the cosine of the imaginary number bi is a real number.

Armed with this formula, we can solve the other "unsolvable" equation introduced at the chapter's beginning, namely $\cos x = 2$. That is, letting $x = bi$, we have

$$2 = \cos x = \cos(bi) = \frac{e^b + e^{-b}}{2} = \frac{(e^b)^2 + 1}{2e^b},$$

or simply $(e^b)^2 - 4(e^b) + 1 = 0$. From the quadratic formula we find $e^b = 2 \pm \sqrt{3}$, so that $b = \ln(2 \pm \sqrt{3})$. Therefore, far from being unsolvable, the equation $\cos x = 2$ has complex solutions $x = bi = i \ln(2 \pm \sqrt{3})$.

In a similar fashion, but starting with the sine series, Euler found

$$\sin(bi) = bi - \frac{(bi)^3}{1 \cdot 2 \cdot 3} + \frac{(bi)^5}{1 \cdot 2 \cdot 3 \cdot 4 \cdot 5} - \cdots$$

$$= i \left[b + \frac{b^3}{1 \cdot 2 \cdot 3} + \frac{b^5}{1 \cdot 2 \cdot 3 \cdot 4 \cdot 5} + \cdots \right] = i \frac{e^b - e^{-b}}{2}.$$

---

[17] Euler, *Opera Omnia*, Ser. 1, Vol. 6, p. 136.

Then, to determine the sine or cosine of an arbitrary complex number, Euler employed the trigonometric identities $\sin(\alpha + \beta) = \sin\alpha\cos\beta + \cos\alpha\sin\beta$ and $\cos(\alpha + \beta) = \cos\alpha\cos\beta - \sin\alpha\sin\beta$ to get:

$$\sin(a + bi) = \sin a \cos(bi) + \cos a \sin(bi) = \sin a\frac{e^b + e^{-b}}{2} + i\cos a\frac{e^b - e^{-b}}{2}$$

and

$$\cos(a + bi) = \cos a\frac{e^b + e^{-b}}{2} - i\sin a\frac{e^b - e^{-b}}{2}.$$

Note that, in using these identities with complex inputs, Euler was operating (again) on faith.

Lingering doubts about such odd-looking results could be assuaged in various ways. For instance, if we write the real number a as the complex number $a + 0 \cdot i$, then the formulas reduce to

$$\sin(a + 0 \cdot i) = \sin a(1) + i\cos a(0) = \sin a$$

and

$$\cos(a + 0 \cdot i) = \cos a(1) - i\sin a(0) = \cos a.$$

Equally encouraging is that:

$$\sin^2(a + bi) + \cos^2(a + bi) =$$

$$\left[\sin a\frac{e^b + e^{-b}}{2} + i\cos a\frac{e^b - e^{-b}}{2}\right]^2 + \left[\cos a\frac{e^b + e^{-b}}{2} - i\sin a\frac{e^b - e^{-b}}{2}\right]^2$$

$$= \sin^2 a\frac{e^{2b} + 2 + e^{-2b}}{4} + 2i\sin a\cos a\frac{e^{2b} - e^{-2b}}{4} - \cos^2 a\frac{e^{2b} - 2 + e^{-2b}}{4}$$

$$+ \cos^2 a\frac{e^{2b} + 2 + e^{-2b}}{4} - 2i\sin a\cos a\frac{e^{2b} - e^{-2b}}{4}$$

$$- \sin^2 a\frac{e^{2b} - 2 + e^{-2b}}{4}$$

$$= \sin^2 a + \cos^2 a = 1.$$

Here the most famous and important identity from (real) trigonometry transfers intact to the complex domain. As a believer in the power of symbols, Euler must have found this comforting indeed.

And what of *logarithms* of complex numbers? This question can be traced back to a controversy between Johann Bernoulli and Gottfried Wilhelm Leibniz from early in the eighteenth century. The challenge at that time was more limited: to ascertain the nature of logarithms of *negative* numbers.

Bernoulli's solution was simple: he believed that $\ln(-x) = \ln x$ for any $x > 0$. To him, this followed from the rules of logarithms, for

$$2\ln(-x) = \ln(-x)^2 = \ln(x^2) = 2\ln x,$$

so that $\ln(-x) = \ln x$. Should anyone remain unconvinced, Bernoulli gave an alternate proof using differential calculus. By the chain rule, it is clear that

$$D_x[\ln(-x)] = \frac{-1}{-x} = \frac{1}{x} = D_x[\ln x],$$

and from these equal derivatives Bernoulli deduced that $\ln(-x) = \ln x$. An immediate corollary is that $\ln(-1) = \ln 1 = 0$, a result with which Johann Bernoulli was perfectly comfortable.

Leibniz would have none of it. He observed that if we let $x = -2$ in the series expansion

$$\ln(1 + x) = x - \frac{x^2}{2} + \frac{x^3}{3} - \frac{x^4}{4} + \cdots,$$

the outcome is $\ln(-1) = -2 - 2 - \frac{8}{3} - 4 - \frac{32}{5} - \cdots$. Consequently $\ln(-1)$, being the sum of infinitely many negative numbers, is surely not 0 as Johann Bernoulli had asserted.

As if to add insult to injury, Leibniz rejected Bernoulli's calculus proof as well. He believed that the differentiation rule $D_x[\ln x] = 1/x$ held only for positive quantities. To apply it to $\ln(-x)$, as Johann did, was impermissible.

These disagreements left the question unresolved, and so it remained until Euler arrived upon the scene. Aware of the Bernoulli/Leibniz controversy, he took it upon himself to find a definitive answer, which he did in a pair of papers from 1747 and 1749.

Euler first addressed the arguments of his illustrious predecessors. He adamantly rejected Leibniz's claim that the differentiation rule for logs was valid only for $x > 0$. If Leibniz were correct, Euler asserted,

> it shatters the foundation of all analysis, which consists principally in the generality of rules and operations which are deemed true, whatever the nature which one supposes for the quantities to which they are applied.[18]

---

[18] Euler, *Opera Omnia*, Ser. 1, Vol. 19, p. 419.

Spoken like a true believer! Euler could not accept a derivative rule of restricted applicability.

Bernoulli did not get off any easier. While endorsing Johann's reasoning that

$$D_x[\ln(-x)] = D_x[\ln x],$$

Euler stressed that equal derivatives do not guarantee equal functions. He observed, for instance, that $D_x[\ln(2x)] = D_x[\ln x]$, but we certainly cannot deduce from this that $2x = x$. The correct conclusion, of course, is that functions with equal derivatives differ by a constant—as with $\ln(2x) = \ln x + \ln 2$. In analogous fashion, it was clear to Euler that

$$\ln(-x) = \ln[(-1)x] = \ln x + \ln(-1).$$

The constant by which $\ln(-x)$ and $\ln x$ differ was the elusive $\ln(-1)$. This was not, as Bernoulli had claimed, equal to zero. It was Euler's job to find it.

This quest turned out to be easy based on what had come before. Simply take logs of both sides of the equation $e^{i\pi} = -1$ to get $\ln(-1) = \ln(e^{i\pi}) = i\pi$. In this way, logs of negatives were finally identified: if $x > 0$, $\ln(-x) = \ln x + i\pi$. Once again, imaginary quantities appeared in surprising places.

But Euler dug much deeper. He raised the question of *how many* (complex) logarithms a non-zero number possesses. Returning to a favorite characterization, he argued that if $y = \ln x$, then

$$x = e^y = \left(1 + \frac{y}{n}\right)^n, \text{ where } n \text{ is an infinite number.}$$

The number x, said Euler, has as many different logarithms as there are different values of y satisfying this equation of infinite degree.

He argued that the quadratic $(1 + y/2)^2 = x$ has two solutions and the cubic $(1 + y/3)^3 = x$ has three. By analogy, if n is an infinite number, there should be infinitely many y for which $(1 + y/n)^n = x$. In short, a non-zero number should have infinitely many logarithms.

This justification of so provocative a claim could hardly convince even Euler's most loyal fans, but he soon established the infinitude of logarithms in the most concrete fashion imaginable: by providing an explicit recipe for generating them.[19]

Given the complex number $a + bi \neq 0$, he again let $c = \sqrt{a^2 + b^2}$ and chose θ so that $\sin \theta = b/c$. Then for each $k = 0, 1, 2, \ldots$, Euler claimed that

$$\ln(a + bi) = \ln c + i(\theta \pm 2\pi k).$$

---

[19] Euler, *Opera Omnia*, Ser. 1, Vol. 6, pp. 134–135.

As proof, one simply exponentiates the right-hand side and applies Euler's identity:

$$e^{\ln c + i(\theta \pm 2\pi k)} = e^{\ln c} \times e^{i(\theta \pm 2\pi k)} = c\left[\cos\left(\theta \pm 2\pi k\right) + i\sin(\theta \pm 2\pi k)\right]$$

$$= c[\cos\theta + i\sin\theta] = c\left[\frac{a}{c} + i\frac{b}{c}\right] = a + bi.$$

It was another triumph. Not only had Euler corrected both Johann Bernoulli and Gottfried Wilhelm Leibniz, but he had discovered the proper definition of complex logs.

We conclude this chapter with a most unexpected consequence of these ideas, one that Euler mentioned in a letter to Christian Goldbach of June, 1746.[20] There he took up what seems a preposterous challenge: to determine the numerical value of i^i—or as he wrote it, of $\sqrt{-1}^{\sqrt{-1}}$. Could anyone find an imaginary power of an imaginary number?

Of course Euler could. In fact, he found infinitely many. For, if $z = i^i$, then $\ln z = i \ln i$. By the characterization above, the logarithms of $i = 0 + 1 \cdot i$ are

$$\ln\sqrt{1} + i\left(\frac{\pi}{2} \pm 2\pi k\right) = i\left(\frac{\pi}{2} \pm 2\pi k\right),$$

and so $\ln z = i\ln i = -\pi/2 \pm 2\pi k$. Therefore

$$i^i = z = e^{\ln z} = e^{-\pi/2} \times e^{\pm 2\pi k},$$

a result that, in Euler's words, "is all the more remarkable because it is real and includes an infinitude of different real values."[21] For instance, if $k = 0$, we have

$$i^i = e^{-\pi/2} = \frac{1}{\sqrt{e^\pi}} \approx 0.20787958,$$

"which seems extraordinary to me," he wrote to Goldbach. Few would quarrel with that assessment.

Complex numbers were here to stay. A concept only dimly understood for its role in solving cubic equations had been legitimized by the discoveries and influence of Leonhard Euler. Without apology or embarrassment, he treated these numbers as equal players upon the mathematical stage and showed how

---

[20] P. F. Fuss, *Correspondance Mathématique et Physique*, Vol. 1, Johnson Reprint, New York, 1968, p. 383.
[21] Euler, *Opera Omnia*, Ser. 1, Vol. 6, pp. 132–133.

to take their roots, logs, sines, and cosines. In so doing, he gave to complex numbers his invaluable imprimatur.

The torch then passed to a group of brilliant successors. The nineteenth century saw the coming of Gauss, Cauchy (1789–1857), Riemann, and Weierstrass (1815–1897), who collectively advanced the frontier in remarkable fashion. Indeed, the hundred years following Euler might well be described as the Century of Complex Variables, a subject for which his work formed so solid a foundation.

Like all great pioneers, he had pointed the way into the unknown.

CHAPTER **6**
Euler and Algebra

Algebra has its roots—so to speak—in equation-solving. Early discoveries can be traced to classical times, but the first flowering of algebra dates to the Islamic mathematicians of the ninth century. In particular, Muhammad ibn-Musa al-Khowârizmî (ca. 825) wrote a treatise on linear and quadratic equations that influenced not only his colleagues but also Renaissance scholars in Europe who rediscovered it six centuries later. Al-Khowârizmî's text ranks alongside Euclid's *Elements* and Euler's *Introductio* as among the most important mathematics books of all time.

We noted in the previous chapter that algebra became a passion in sixteenth-century Italy. Its history resounds with names like Scipione del Ferro, Girolamo Cardano, and Rafael Bombelli. But for all of the achievements of these Italian algebraists, theirs was an algebra without symbols. Lacking notation, they provided intricate verbal recipes for solving equations, a procedure that seems impossibly cumbersome to us today. Algebraic symbolism had to await the 1591 publication of *The Analytic Art* by François Viète (1540–1603).

In algebra, as in so many branches of mathematics, the seventeenth century saw decisive progress. Descartes's *Géométrie* from 1639, for instance, contained equations written in essentially modern form. By Euler's arrival a century later, algebraic notation had become part of the fabric of mathematics. Not surprisingly, that great mathematical expositor wrote a textbook on the subject, his *Elements of Algebra*. There Euler succinctly defined algebra as "the science which teaches how to determine unknown quantities by means of those that are known."[1]

Al-Khowârizmî, Cardano, Viète, Euler—these are some of the authors who created elementary algebra. This is not to say that they resolved every issue satisfactorily. In Euler's day two major algebraic questions remained

---

[1] Euler, *Elements of Algebra*, p. 186.

open, and both will concern us in this chapter. One was the problem of solving polynomial equations of arbitrary degree. The other was what we now call, with obvious reverence, *the* fundamental theorem of algebra.

Euler made contributions to both, although in neither case was his work ultimately decisive. For different reasons, one could say that he "failed" in each. But his investigations were so clever, his insights so penetrating, that they deserve our attention.

Prologue

In Chapter 5, we saw how the work of del Ferro and Cardano led directly to an algebraic solution of the cubic and indirectly to a recognition of complex numbers. But the Italian algebraists did not stop there. Around 1540 Ludovico Ferrari (1522–1565), Cardano's friend and protégé, described a solution by radicals of the quartic, or fourth-degree, equation. Ferrari's technique was first published in Chapter 39 of the *Ars Magna*, where Cardano generously gave credit to its discoverer. Of course, it had to be presented entirely in verbal form, no simple matter for so complicated a procedure. Mathematicians of the next century—in particular Descartes and Newton—had the advantage of algebraic symbolism when solving quartics.[2]

By the time of Euler, this discovery was two centuries old. But, as he demonstrated repeatedly throughout his long career, one can seek new routes to familiar destinations. In his *Elements of Algebra*, he described a method for solving fourth-degree equations "altogether different" from what had come before.[3] To give a sense of Euler's view of algebra, not to mention the intricacies of equation-solving from the past, we shall examine his solution in detail.

Suppose we have a general quartic equation $Ay^4 + By^3 + Cy^2 + Dy + E = 0$. "We must begin," Euler observed, "by destroying the second term." This slightly violent expression meant only that we replace the quartic by a related one lacking its cubic term. This is accomplished by dividing by A, making the substitution $y = x - B/4A$, collecting terms, and simplifying. The result will have the form (in Euler's notation) $x^4 - ax^2 - bx - c = 0$. This is a so-called "depressed quartic," the obvious counterpart of the depressed cubic we saw in the previous chapter.

---

[2] See Descartes, pp. 180–187 and Whiteside, *The Mathematical Papers of Isaac Newton*, Vol. 5, p. 413.

[3] Euler, *Elements of Algebra*, pp. 282–288.

Euler now assumed that a solution to the depressed quartic takes the form

$$x = \sqrt{p} + \sqrt{q} + \sqrt{r}, \tag{6.1}$$

where the unknowns p, q, and r must be determined in terms of a, b, and c.

To do so, Euler squared (6.1) and simplified to get

$$x^2 - (p + q + r) = 2\left(\sqrt{pq} + \sqrt{pr} + \sqrt{qr}\right).$$

Upon squaring both sides again, he arrived at:

$$\begin{aligned}
x^4 &- 2(p + q + r)x^2 + (p + q + r)^2 \\
&= 4(pq + pr + qr) + 8\left(\sqrt{p^2qr} + \sqrt{pq^2r} + \sqrt{pqr^2}\right) \\
&= 4(pq + pr + qr) + 8\sqrt{pqr}\left(\sqrt{p} + \sqrt{q} + \sqrt{r}\right) \\
&= 4(pq + pr + qr) + 8\sqrt{pqr}\,x.
\end{aligned} \tag{6.2}$$

Euler next introduced auxiliary variables

$$f = p + q + r, \quad g = pq + pr + qr, \quad \text{and} \quad h = pqr. \tag{6.3}$$

Equation (6.2) was thereby recast as $x^4 - 2fx^2 - 8\sqrt{h}x - (4g - f^2) = 0$, which, when compared to the original depressed quartic $x^4 - ax^2 - bx - c = 0$, revealed that:

(a) $2f = a$, and so $f = a/2$,
(b) $8\sqrt{h} = b$, and so $h = b^2/64$,

and

(c) $4g - f^2 = c$, and so $g = (4c + a^2)/16$.

In this way, Euler had related the coefficients of the depressed quartic to the auxiliary quantities f, g, and h. Of course, his true objective was to relate these coefficients to p, q, and r, and the secret to doing so was contained in equations (6.3).

That is, Euler recognized p, q, and r as the roots of

$$\begin{aligned}
0 = (z - p)(z - q)(z - r) &= z^3 - (p + q + r)z^2 + (pq + pr + qr)z - pqr \\
&= z^3 - fz^2 + gz - h.
\end{aligned}$$

And here at last is the key to the puzzle. From the known values of a, b, and c, we find f, g, and h. From these, we construct the cubic $z^3 - fz^2 + gz - h = 0$, the solutions to which—found via Cardano's formula—are $z = p$, $z = q$, and $z = r$. Finally, the solution of the depressed quartic is $x = \sqrt{p} + \sqrt{q} + \sqrt{r}$, and the solution of the original (non-depressed) quartic follows from $y = x - B/4A$.

As an example of this procedure, we consider a problem taken from Euler: solve the quartic $y^4 - 8y^3 + 14y^2 + 4y - 8 = 0$.

The first step is to depress it by means of $y = x - \frac{-8}{4} = x + 2$. After we substitute, the quartic becomes $x^4 - 10x^2 - 4x + 8 = 0$. We then identify

$$f = \frac{a}{2} = \frac{10}{2} = 5, \quad g = \frac{4c + a^2}{16} = \frac{4(-8) + 100}{16} = \frac{17}{4}, \quad \text{and}$$

$$h = \frac{b^2}{64} = \frac{4^2}{64} = \frac{1}{4}.$$

Thus p, q, and r are the solutions of the auxiliary *cubic* $z^3 - 5z^2 + \frac{17}{4}z - \frac{1}{4} = 0$.

Finding this unwieldy, Euler substituted $z = u/2$ and simplified to get the more manageable equivalent:

$$u^3 - 10u^2 + 17u - 2 = 0.$$

Theoretically, this could be solved via Cardano's technique, but Euler (with the great good fortune accorded only to textbook writers) noticed that $u = 2$ is an obvious solution. The cubic therefore factors as

$$0 = u^3 - 10u^2 + 17u - 2 = (u - 2)(u^2 - 8u + 1),$$

whose three solutions are $u = 2$, $u = 4 + \sqrt{15}$, and $u = 4 - \sqrt{15}$.

Now all of this must be unraveled.

First, because $z = u/2$, the solutions of the cubic in z are $p = 1$, $q = (4 + \sqrt{15})/2$, and $r = (4 - \sqrt{15})/2$, and it follows that

$$\sqrt{p} = \pm 1, \quad \sqrt{q} = \pm\frac{1}{2}\sqrt{8 + 2\sqrt{15}}, \quad \text{and} \quad \sqrt{r} = \pm\frac{1}{2}\sqrt{8 - 2\sqrt{15}}.$$

Euler's sharp eye recognized something more: as is easily checked, the term $8 + 2\sqrt{15}$ can be written as $(\sqrt{5} + \sqrt{3})(\sqrt{5} + \sqrt{3})$. Consequently, $\sqrt{8 + 2\sqrt{15}} = \sqrt{5} + \sqrt{3}$. In like manner, it is clear that $\sqrt{8 - 2\sqrt{15}} = \sqrt{5} - \sqrt{3}$. This allows the further simplification:

$$\sqrt{p} = \pm 1, \quad \sqrt{q} = \pm\frac{\sqrt{5} + \sqrt{3}}{2}, \quad \text{and} \quad \sqrt{r} = \pm\frac{\sqrt{5} - \sqrt{3}}{2}.$$

Moving toward a conclusion, Euler observed that $b/8 = \sqrt{h} = \sqrt{pqr} = \sqrt{p} \times \sqrt{q} \times \sqrt{r}$, and because in this example $b = 4 > 0$, signs must be attached to the three square roots so their product is positive. The depressed quartic thus has the four solutions:

$$x_1 = \sqrt{p} + \sqrt{q} + \sqrt{r} = 1 + \frac{\sqrt{5} + \sqrt{3}}{2} + \frac{\sqrt{5} - \sqrt{3}}{2} = 1 + \sqrt{5};$$

$$x_2 = \sqrt{p} - \sqrt{q} - \sqrt{r} = 1 - \frac{\sqrt{5} + \sqrt{3}}{2} - \frac{\sqrt{5} - \sqrt{3}}{2} = 1 - \sqrt{5};$$

$$x_3 = -\sqrt{p} + \sqrt{q} - \sqrt{r} = -1 + \frac{\sqrt{5} + \sqrt{3}}{2} - \frac{\sqrt{5} - \sqrt{3}}{2} = -1 + \sqrt{3};$$

$$x_4 = -\sqrt{p} - \sqrt{q} + \sqrt{r} = -1 - \frac{\sqrt{5} + \sqrt{3}}{2} + \frac{\sqrt{5} - \sqrt{3}}{2} = -1 - \sqrt{3}.$$

Finally (!), the solutions of the *original* quartic $y^4 - 8y^3 + 14y^2 + 4y - 8 = 0$ are determined (from the relationship $y = x + 2$) to be $y = 3 \pm \sqrt{5}$ and $y = 1 \pm \sqrt{3}$. Needless to say, all of these check.

One aspect of this problem—apart from its length—deserves particular note: solving the quartic rests upon the ability to solve a related *cubic*. This echoes the solution of the cubic which, as we saw in Chapter 5, required the solution of a related *quadratic*.

It thus seems likely that, in order to solve the quintic (fifth-degree) equation, one should: first transform it into a depressed quintic; introduce auxiliary variables whose values are determined from a related quartic; solve this quartic by the techniques just described; and, once done, reassemble the solutions of the quartic to get the solutions of the quintic.

Euler, however, provided no such development. On the contrary, his treatment of the quartic was followed not by a solution of the quintic but by these words:

This is the greatest length to which we have yet arrived in the resolution of algebraic equations. All the pains that have been taken in order to resolve equations of the fifth degree, and those of higher dimensions, in the same manner, or, at least, to reduce them to inferior degrees, have been unsuccessful; so that we cannot give any general rules for finding the roots of equations which exceed the fourth degree.[4]

Euler, to put it bluntly, was stuck. He certainly would have coveted the distinction of being the first mathematician to solve the quintic, and one senses that

---

[4]Ibid., p. 286.

mere algebraic complexity would not have stopped him. But he came up empty. Solving the general polynomial equation is the first of the great algebraic questions, unanswered in Euler's day, mentioned in this chapter's introduction. We shall return to this matter in the epilogue.

With regard to the second question—the fundamental theorem of algebra—Euler had more to say. In the eighteenth century, this result was framed in terms of the ultimate factorization of a real polynomial. That is, mathematicians conjectured that every real polynomial can be expressed as the product of real linear and/or real quadratic factors.

As an example, we consider the factorization:

$$3x^5 + 5x^4 + 10x^3 + 20x^2 - 8x = x(3x - 1)(x + 2)(x^2 + 4).$$

Here a quintic has been shattered into the product of three linear pieces and one irreducible quadratic one, and all polynomials in sight are real. The conjecture asserted that such a factorization *exists* for any real polynomial, no matter its degree. We stress that this was a pure existence statement. It did not provide an explicit formula for the various factors.

Anticipating a bit, we see that we can further factor irreducible quadratics if we operate within the complex numbers. For our example, we have

$$3x^5 + 5x^4 + 10x^3 + 20x^2 - 8x = x(3x - 1)(x + 2)(x - 2i)(x + 2i).$$

This is "complete" in the sense that a real fifth-degree polynomial has been factored into the product of five *linear* complex factors, certainly as far as any decomposition can hope to proceed. It is in this light—the factorization of an nth degree polynomial into n linear factors—that the fundamental theorem of algebra is now perceived.

There were good reasons for our mathematical predecessors to accept this conjecture as plausible. For instance, recall Euler's discussion of complex roots from Chapter 5. We saw that the polynomial equation $x^n - 1 = 0$ has n complex solutions—the nth roots of unity—given by:

$$\omega_k = \cos\frac{2\pi k}{n} + i\sin\frac{2\pi k}{n} \quad \text{for } k = 0, 1, 2, \ldots, n - 1.$$

This immediately provides a decomposition of $x^n - 1$ into n linear, complex factors:

$$x^n - 1 = (x - \omega_0)(x - \omega_1)(x - \omega_2)\cdots(x - \omega_{n-1}).$$

Granted, $x^n - 1$ is a very special nth degree polynomial, but perhaps a similar phenomenon occurs for its more intricate brethren.

On the other hand, there were skeptics. No less an authority than Leibniz doubted that every real polynomial can be factored into real linear and/or real quadratic pieces.[5] Worse, Nicolaus Bernoulli (1687–1759) claimed to have found a counterexample—namely, $x^4 - 4x^3 + 2x^2 + 4x + 4$—that could not be so factored. If he were correct, the game was over: the fundamental "theorem" of algebra would have been automatically disproved.

Euler, rising to the defense of this conjecture, showed that Bernoulli was wrong. In a 1742 letter to Christian Goldbach (during which Euler fell into the annoying habit of switching between German and Latin in mid-sentence), he factored the supposedly unfactorable, splitting the quartic into the product of quadratics[6]

$$x^2 - \left(2 + \sqrt{4 + 2\sqrt{7}}\right) x + \left(1 + \sqrt{4 + 2\sqrt{7}} + \sqrt{7}\right)$$

and

$$x^2 - \left(2 - \sqrt{4 + 2\sqrt{7}}\right) x + \left(1 - \sqrt{4 + 2\sqrt{7}} + \sqrt{7}\right).$$

This factorization appears to lie somewhere between the miraculous and the preposterous. It looks ever so much like a misprint—but it is perfectly correct. Anyone with a taste for computation can verify that these two complicated, root-infested factors multiply to yield Bernoulli's simple quartic. Far more challenging, of course, is to figure out *how* Euler derived this factorization in the first place. (Hint: it was not by guessing).

So the purported counterexample was no such thing, and the conjecture remained viable. One who enthusiastically endorsed it was Jean d'Alembert (1717–1783). In 1746 he offered a proof.[7] For d'Alembert, the theorem had a dual significance—it not only addressed a fundamental issue in algebra but also resolved an important problem in integral calculus.

As an illustration of the latter, consider the indefinite integral

$$\int \frac{34x^4 + 6x^3 + 89x^2 + 26x - 16}{3x^5 + 5x^4 + 10x^3 + 20x^2 - 8x}\, dx.$$

Needless to say, this does not appear in any integral table. Indeed, it gives a workout even to computer packages featuring symbolic manipulation (unavailable to eighteenth-century mathematicians in any case).

---

[5] Kline, p. 597.

[6] Fuss, Vol. 1, pp. 170–171.

[7] Dirk Struik, ed., *A Source Book in Mathematics: 1200–1800*, Princeton U. Press, 1986, p. 99.

But, as we noted above, the denominator of the integrand can be factored to give

$$\int \frac{34x^4 + 6x^3 + 89x^2 + 26x - 16}{x(3x - 1)(x + 2)(x^2 + 4)}\, dx.$$

We then determine the partial fraction decomposition and integrate the pieces in terms of elementary functions:

$$\int \frac{34x^4 + 6x^3 + 89x^2 + 26x - 16}{x(3x - 1)(x + 2)(x^2 + 4)}\, dx$$

$$= \int \frac{2}{x}\, dx + \int \frac{1}{3x - 1}\, dx + \int \frac{7}{x + 2}\, dx + \int \frac{2x - 3}{x^2 + 4}\, dx$$

$$= 2 \ln |x| + \frac{1}{3} \ln |3x - 1| + 7 \ln |x + 2|$$

$$+ \ln(x^2 + 4) - \frac{3}{2} \tan^{-1}(x/2) + C.$$

Although far from self-evident, it can be checked that this expression is indeed the antiderivative of

$$\frac{34x^4 + 6x^3 + 89x^2 + 26x - 16}{3x^5 + 5x^4 + 10x^3 + 20x^2 - 8x}.$$

Were the fundamental theorem of algebra proved in general, it would follow that for *any* rational function $P(x)/Q(x)$ where P and Q are real polynomials, the indefinite integral $\int (P(x)/Q(x))\, dx$ *exists* as a combination of fairly simple functions. We need only perform long division to reduce this rational expression to one where the degree of the numerator is less than the degree of $Q(x)$; next we consider $Q(x)$ as the product of its real linear and/or real quadratic factors; then we apply the partial fraction technique to break the integral into components of the form

$$\int \frac{A}{(ax + b)^n}\, dx \quad \text{and/or} \quad \int \frac{Bx + C}{(ax^2 + bx + c)^n}\, dx;$$

and finally we determine these indefinite integrals using nothing more complicated than natural logarithms, inverse tangents, or trigonometric substitution. Euler quite properly called this a "beautiful and important consequence" of the fundamental theorem.[8]

---

[8] Euler, *Opera Omnia*, Ser. 1, Vol. 6, p. 107.

As noted previously, the fundamental theorem is not accompanied by an algorithm for finding the denominator's factors; but, just as it guarantees the *existence* of such a factorization, so too will the *existence* of simple antiderivatives for any rational function be established.

Unfortunately, d'Alembert's attempt to prove the theorem was unsuccessful, for the difficulties it presented were simply too great for him to overcome.[9] Consequently, as of 1746, the situation remained unclear. Mathematicians were faced with a proposition of great importance to both algebra and analysis, yet its validity was by no means certain. Someone else had to step forward to give it a try.

Enter Euler

Euler believed in the conjecture. As early as 1742 he declared to Goldbach that "All algebraic expressions $\alpha + \beta x + \gamma x^2 + \delta x^3 + \epsilon x^4 +$ etc. can be resolved either into simple real factors $p + qx$ or else into real quadratic factors $p + qx + rx^2$."[10] Later, in the *Introductio*, he wrote, "If there is any doubt that every polynomial can be expressed as a product of real linear and real quadratic factors, then that doubt by this time should be almost completely dissipated."[11]

Of course, "almost completely dissipated" is not the same as "proved." And so, in 1749 Euler presented his own demonstration of the general result. It was part of his paper *"Recherches sur les racines imaginaires des équations"* mentioned in Chapter 5. We stress at the outset that he was unsuccessful in proving the fundamental theorem of algebra. As we shall see, Euler's reasoning suffered logical shortcomings. Even so, one cannot fail to recognize the deftness of a master at work.

Rather than attack the general polynomial directly, Euler began with simple cases and worked toward more difficult ones (almost always a wise course of action). First he addressed the quartic.[12]

Theorem. *Any quartic polynomial $x^4 + Ax^3 + Bx^2 + Cx + D$ where A, B, C, and D are real can be decomposed into two real factors of the second degree.*

Proof. Using the standard opening gambit, Euler first substituted $x = y - A/4$ to depress the quartic. There were advantages to factoring a depressed

---

[9] See John Stillwell, *Mathematics and its History*, Springer-Verlag, New York, 1989, pp. 195–200.
[10] Fuss, Vol. 1, p. 171.
[11] Euler, *Introduction to Analysis of the Infinite*, Book I, p. 124.
[12] Euler, *Opera Omnia*, Ser. 1, Vol. 6, pp. 93–94.

quartic rather than a full-blown one, yet a factorization of the former yields a factorization of the latter via the reverse substitution $y = x + A/4$.

It was thus sufficient for him to consider $x^4 + Bx^2 + Cx + D$, where B, C, and D are real. At this juncture, two cases presented themselves.

Case 1: C = 0.

Here we have a quartic $x^4 + Bx^2 + D$, which is a quadratic in x^2. A pair of sub-cases arise:

First, if $B^2 - 4D \geq 0$, we apply the quadratic formula to get a decomposition into two second-degree *real* factors:

$$x^4 + Bx^2 + D = \left[x^2 + \frac{B - \sqrt{B^2 - 4D}}{2} \right] \left[x^2 + \frac{B + \sqrt{B^2 - 4D}}{2} \right].$$

For instance, $x^4 + x^2 - 12 = (x^2 - 3)(x^2 + 4)$.

Less direct is the subcase where we factor $x^4 + Bx^2 + D$ when $B^2 - 4D < 0$. The previous decomposition no longer works because the factors containing $\sqrt{B^2 - 4D}$ are not real. However, the quartic can be re-written as the difference of squares and factored as:

$$x^4 + Bx^2 + D = \left[x^2 + \sqrt{D} \right]^2 - \left[x\sqrt{2\sqrt{D} - B} \right]^2$$

$$= \left[x^2 + \sqrt{D} - x\sqrt{2\sqrt{D} - B} \right] \left[x^2 + \sqrt{D} + x\sqrt{2\sqrt{D} - B} \right].$$

A few points must be made about this result. First, the condition $B^2 - 4D < 0$ implies that $4D > B^2 \geq 0$, and so the expression $\sqrt{D}$ in the preceding factorization is real. Likewise, $4D > B^2$ guarantees that $2\sqrt{D} > |B| \geq B$, and so $\sqrt{2\sqrt{D} - B}$ is a real number as well. In short, the factors above are two real quadratics.

For example, when factoring $x^4 + x^2 + 4$, we have $B^2 - 4D = -15 < 0$, and our procedure yields $x^4 + x^2 + 4 = [x^2 - x\sqrt{3} + 2][x^2 + x\sqrt{3} + 2]$.

Case 2: C ≠ 0.

This is the more difficult scenario. Euler observed that any factorization of the depressed quartic into real quadratics *must* take the form

$$x^4 + Bx^2 + Cx + D = (x^2 + ux + \alpha)(x^2 - ux + \beta) \qquad (6.4)$$

for real numbers u, α, and β yet to be determined. This is necessary because the "ux" in one factor must have a compensating "$-ux$" in the other to obliterate the cubic term.

Euler expanded the right-hand side of (6.4) to get

$$x^4 + Bx^2 + Cx + D = x^4 + (\alpha + \beta - u^2)x^2 + (\beta u - \alpha u)x + \alpha\beta$$

and matched coefficients of like powers to generate the three equations:

$$B = \alpha + \beta - u^2, \quad C = \beta u - \alpha u = (\beta - \alpha)u, \quad \text{and} \quad D = \alpha\beta.$$

Recall that B, C, and D are the known coefficients of the depressed quartic, whereas u, α, and β are unknown real numbers whose *existence* Euler had to establish.

From the first two equations he concluded that

$$\alpha + \beta = B + u^2 \quad \text{and} \quad \beta - \alpha = \frac{C}{u}.$$

As an important aside, we note that, because $0 \neq C = (\beta - \alpha)u$, then u itself is non-zero, so its presence in the denominator above is no cause for alarm.

Euler next added and subtracted these two equations to get

$$2\beta = B + u^2 + \frac{C}{u} \quad \text{and} \quad 2\alpha = B + u^2 - \frac{C}{u}. \tag{6.5}$$

But $D = \alpha\beta$, and so

$$4D = 4\alpha\beta = (2\beta)(2\alpha)$$

$$= \left(B + u^2 + \frac{C}{u}\right)\left(B + u^2 - \frac{C}{u}\right) = u^4 + 2Bu^2 + B^2 - \frac{C^2}{u^2}.$$

Finally, multiplying this equation by u^2 and simplifying gave him:

$$u^6 + 2Bu^4 + (B^2 - 4D)u^2 - C^2 = 0. \tag{6.6}$$

It may appear that things have gotten worse, for Euler had traded a fourth-degree equation in x for a sixth-degree equation in u. Admittedly, (6.6) is also a cubic in u^2, and because every real cubic has a real solution, one could properly conclude that there is a real value for u^2 satisfying (6.6). However, a moment's reflection shows that this does not guarantee the existence of a *real* value for u (e.g., if $u^2 = -1$), which was Euler's true objective.

Undeterred, he exploited four critical properties of (6.6):

(a) B, C, and D are given, so the only unknown here is u.

(b) B, C, and D are real.

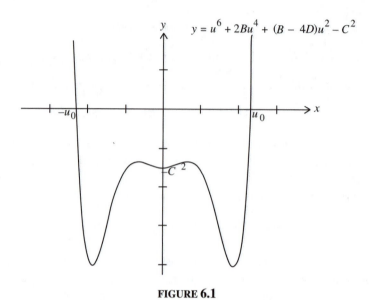

$$y = u^6 + 2Bu^4 + (B - 4D)u^2 - C^2$$

FIGURE 6.1

(c) the polynomial is even, and thus its graph is symmetric about the y-axis.

(d) the constant term of this sixth-degree polynomial is $-C^2 < 0$.

Euler had therefore generated a sixth-degree real polynomial in u whose graph looks something like that shown in Figure 6.1. It has a negative y-intercept at $(0, -C^2)$ because C is a non-zero real number, and its graph climbs toward $+\infty$ as u becomes unbounded in either the positive or negative direction. By the polynomial's continuity and the intermediate value theorem—which Euler took as intuitively clear—he was guaranteed the *existence* of real numbers $u_0 > 0$ and $-u_0 < 0$ satisfying this sixth-degree polynomial.

Using the positive solution u_0, Euler solved for β and α in (6.5), getting real solutions

$$\beta_0 = \frac{1}{2}\left(B + u_0^2 + \frac{C}{u_0}\right) \quad \text{and} \quad \alpha_0 = \frac{1}{2}\left(B + u_0^2 - \frac{C}{u_0}\right),$$

and, because $u_0 > 0$, these fractions are well-defined.

To summarize, in the case that $C \neq 0$, Euler had established the existence of real numbers u_0, α_0, and β_0 such that

$$x^4 + Bx^2 + Cx + D = (x^2 + u_0 x + \alpha_0)(x^2 - u_0 x + \beta_0).$$

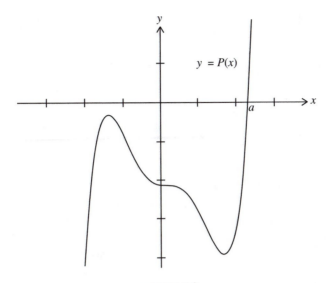

FIGURE 6.2

He thereby had proved that any depressed quartic with real coefficients—and by extension any real quartic at all—can be factored into two real quadratics, whether or not $C = 0$.   Q.E.D.

At this point Euler observed, "it is also evident that any equation of the fifth degree is resolvable into three real factors of which one is linear and two are quadratic."[13] His reasoning was simple (see Figure 6.2). An odd-degree polynomial—and consequently a fifth-degree polynomial $P(x)$—is guaranteed by the intermediate value theorem to have at least one real x-intercept, say at $x = a$. Thus $P(x) = (x - a)Q(x)$, where $Q(x)$ is a polynomial of the fourth degree, and the previous result allows us to decompose $Q(x)$ into two real quadratic factors.

By now, a strategy was brewing in Euler's mind. Rather than considering polynomials of degree 6, 7, 8, and so on, he saw a way to simplify the challenge. He realized that *if* he could prove his decomposition for real polynomials of degree 4, 8, 16, 32, and in general of degree 2^n, then he could prove it for any real polynomials whatever.

---

[13] Ibid., p. 95.

Why is this? Suppose, for instance, Euler wished to establish that the polynomial

$$x^{12} - 3x^9 + 52x^8 + 3x^3 - 2x + 17$$

can be written as the product of real linear and/or real quadratic factors. He would multiply by x^4 to get

$$x^{16} - 3x^{13} + 52x^{12} + 3x^7 - 2x^5 + 17x^4,$$

and then—assuming the result had been proved for degree 16—he would know that this polynomial had such a factorization, obviously containing the four linear factors x, x, x, and x. By cancelling these, he would of necessity be left with real linear and/or real quadratic factors of the original 12th-degree polynomial.

Adopting this strategy, Euler's next objective was to prove "Any equation of the eighth degree is always resolvable into two real factors of the fourth degree."[14] Because each of the fourth-degree factors was itself decomposable into a pair of real quadratics, which themselves can be broken into (possibly complex) linear factors, he would have succeeded in shattering the eighth-degree polynomial into eight linear pieces.

Unfortunately, an analogous attack on the octic polynomial becomes exceedingly complicated. One first depresses the octic and considers a factorization into a pair of quartics:

$$\begin{aligned}
&x^8 + Bx^6 + Cx^5 + Dx^4 + Ex^3 + Fx^2 + Gx + H \\
&= (x^4 + ux^3 + \alpha x^2 + \beta x + \gamma)(x^4 - ux^3 + \delta x^2 + \epsilon x + \phi).
\end{aligned} \tag{6.7}$$

One multiplies the quartics, equates the resulting coefficients with the known quantities $B, C, D, \ldots$ to get seven equations in seven unknowns, and asserts that there exist real values of $u, \alpha, \beta, \gamma, \ldots$ satisfying this system.

The parallels with the previous case are evident. But what made this so unsatisfactory was Euler's admission that

> when I pass to equations of very high degree, it will be very difficult and even impossible to find the equation by which the unknown u is determined.

He was, in short, unable to solve this system explicitly for u. The proof had collapsed.

---

[14]Ibid., p. 99.

Ever resourceful, Euler decided to look again for inspiration to the depressed quartic in (6.4). An entirely different line of reasoning suggested itself, a line he thought could be extended naturally and successfully to the eighth and higher-degree cases.[15]

He began by *assuming* that the quartic in (6.4) has four roots p, q, r, and s, so that

$$(x^2 + ux + \alpha)(x^2 - ux + \beta) = x^4 + Bx^2 + Cx + D$$
$$= (x - p)(x - q)(x - r)(x - s).$$
(6.8)

From this hypothesized factorization he drew three conclusions.

First, upon multiplying the four linear factors on the right of (6.8), we see that the coefficient of x^3 is $-(p + q + r + s)$. Hence $p + q + r + s = 0$ because the quartic is depressed.

Second, the quadratic factor $(x^2 - ux + \beta)$ must arise as the product of two of the four linear factors. That is, $(x^2 - ux + \beta)$ could be $(x - p)(x - r) = x^2 - (p + r)x + pr$; it could just as well be $(x - q)(x - r) = x^2 - (q + r)x + qr$; and so on. This implied that, in the first case, $u = p + r$, whereas in the second $u = q + r$. In fact, it was clear to Euler that u can take any of the $\binom{4}{2} = 6$ values:

$$R_1 = p + q \qquad R_4 = r + s$$
$$R_2 = p + r \qquad R_5 = q + s$$
$$R_3 = p + s \qquad R_6 = q + r.$$

Because u is an unknown having these six possible values, it must be determined by the sixth-degree polynomial

$$(u - R_1)(u - R_2)(u - R_3)(u - R_4)(u - R_5)(u - R_6).$$

This situation, of course, is consistent with the sixth-degree polynomial in u that Euler had found in (6.6).

He made one additional observation. From $p + q + r + s = 0$, it follows that $R_4 = -R_1$, $R_5 = -R_2$, and $R_6 = -R_3$. Hence the sixth-degree polynomial becomes

$$(u - R_1)(u + R_1)(u - R_2)(u + R_2)(u - R_3)(u + R_3)$$
$$= (u^2 - R_1^2)(u^2 - R_2^2)(u^2 - R_3^2).$$

---

[15] Ibid., pp. 96–106.

The constant term here—and hence this polynomial's y-intercept—is

$$-R_1^2 R_2^2 R_3^2 = -(R_1 R_2 R_3)^2.$$

This constant, Euler claimed, was a negative real number, again in complete agreement with his conclusions from equation (6.6).

To summarize, Euler had provided an entirely different argument to establish that, in the quartic case, u is determined by a polynomial of degree $\binom{4}{2} = 6$ with a negative y-intercept. This was the critical conclusion he had already drawn—and from which the result follows as above—but here he drew it without *explicitly* finding the equation determining u.

The advantage of this second approach to the quartic case was that it could be applied to the depressed octic as well. *Assuming* that the octic of (6.7) was decomposed into eight linear factors, Euler mimicked the reasoning above to deduce that, for each different combination of four of these eight factors, we get a different value of u. Then u would be determined by a polynomial of degree $\binom{8}{4} = 70$ having a negative y-intercept. He next employed the intermediate value theorem to assert the existence of a *real* root u_0, and from this he deduced that the other real numbers α_0, β_0, γ_0, δ_0, ϵ_0, and ϕ_0 exist as well.

Euler argued similarly in the 16th-degree case, claiming that "... the equation which determines the values of the unknown u will necessarily be of the 12,870th degree."[16] That is, the degree of this (obviously unspecified) equation is $\binom{16}{8} = 12{,}870$. Clearly, Euler's observation that it would be "very difficult and even impossible" to find these polynomials explicitly had become something of an understatement.

From there it was a short and entirely analogous step to the general case: that any real polynomial of degree 2^n could be factored into two real polynomials of degree 2^{n-1}. With that, Euler's proof was finished.

Or was it?

Unfortunately, his treatment of the 8th-degree, 16th-degree, and general cases contained logical holes. For instance, if we look back at the quartic, how could Euler assert that it has four roots? How could he assert that the octic has eight?

More significantly, what is the *nature* of these supposed roots? Are they real? Are they complex? Or are they an unspecified—and perhaps never before encountered—kind of number? If so, can they be added and multiplied in the usual fashion?

---

[16]Ibid., p. 103.

These are not incidental questions. In the quartic of (6.8), for example, if we are uncertain about the nature of the roots p, q, r, and s, then we are equally uncertain about the nature of their sums R_1, R_2, R_3. Consequently, there is no guarantee whatever that the expression $-(R_1 R_2 R_3)^2$ is a negative real number, and if this y-intercept is not a negative real, the intermediate value argument cannot be invoked.

It appears, then, that Euler had started down a promising road in his pursuit of the fundamental theorem. His first proof worked nicely in dealing with fourth- and fifth-degree real polynomials. But as he pursued this elusive theorem deeper into the thicket, complications involving the existence of his desired real factors became overwhelming. In a certain sense, he lost his way among the high-degree polynomials that beckoned him on, and his general proof vanished in the wilderness.

The theorem was as yet unproved. So it would remain for half a century until another mathematician, standing upon Euler's shoulders, would be able to see what Euler had not.

Epilogue

In the interest of historical completeness, we shall use this epilogue to describe what became of the two algebraic questions of this chapter, questions for which Euler's contributions were inconclusive.

The first was a solution by radicals of equations with degree higher than the fourth. Mathematicians in the generation after Euler were equally unsuccessful in solving the quintic algebraically. Finally, long after Euler conceded that "we cannot give any general rules for finding the roots of equations which exceed the fourth degree," the matter was settled in the negative: a solution by radicals of the general fifth-degree equation was proved to be impossible.

It was the Norwegian Niels Abel (1802–1829) who established this result in papers of 1824 and 1826.[17] Abel, whose premature death deprived the world of an extraordinarily gifted mathematician, proved that the quintic is not solvable by radicals. By this we do not mean that its solution is "difficult" or "as yet undiscovered" or "beyond the current limits of mathematical knowledge." Rather, he demonstrated that for the general fifth-degree equation there *cannot* be a formal solution involving only the equation's coefficients and the algebraic operations of addition, subtraction, multiplication, division, and root extraction. The quadratic formula has no counterpart of degree five or higher.

---

[17] Smith, pp. 261–266.

Abel's proof is far from simple and will not be explained here. Suffice it to say that, upon assuming the general quintic to be solvable, he eventually reached the contradiction that a certain expression, when viewed in one light, takes exactly five different values but, when viewed in another, takes 5! = 120 different values. Such an absurdity, when traced back through the logical machinery of his argument, allowed Abel to conclude that the initial assumption of solvability was untenable.

So, the quest had been futile all along. A solution by radicals of the quintic—or of any higher-degree equation—was as impossible as trying to find a fraction whose square is 2. The standard algebraic operations are just not powerful enough to solve any equations but those of low degree.

Considerations of this type certainly demonstrated the limitations of algebra, but they also led to a new and deeper understanding of the subject. One can trace the origin of concepts like "group" and "field"—cornerstones of modern abstract algebra—to concrete problems such as this from the nineteenth century.[18]

Incidentally, we observe that Abel's discovery had one other (admittedly minor) consequence: it let Euler off the hook for his failure to solve the quintic. We can hardly fault someone for not doing the impossible.

Euler fares less well regarding the fundamental theorem of algebra, for a proof was forthcoming within two decades of his death.[19] It appeared in the 1799 doctoral dissertation of Carl Friedrich Gauss, a treatise with the rambling title, "A New Proof of the Theorem That Every Integral Rational Algebraic Function [i.e., every polynomial with real coefficients] Can Be Decomposed into Real Factors of the First or Second Degree."[20]

Gauss began his thesis with a critique of past attempts at a proof. When addressing Euler's argument, he raised the issues cited above, designating Euler's mysterious, hypothesized roots as "shadowy." To Gauss, Euler's attempt lacked "the clarity which is required in mathematics."[21] This clarity he attempted to provide, not only in the dissertation but in proofs from 1815, 1816, and 1848.

These days, the fundamental theorem of algebra tends to be proved in a complex *analysis* course. There, we consider functions mapping complex numbers to complex numbers and examine their analytic properties—e.g.,

---

[18]See, for instance, Israel Kleiner, "The Teaching of Abstract Algebra: An Historical Perspective," in *Learn from the Masters*, Mathematical Association of America, Washington, D.C., 1995, pp. 225–239.

[19]Stillwell, p. 196, offers an interesting twist on this oft-repeated statement.

[20]Struik, pp. 115–122.

[21]Fauvel and Gray, p. 491.

boundedness, differentiability, and integrability. As noted in the previous chapter, such matters became the passion of nineteenth-century mathematicians like Cauchy, Riemann, and Weierstrass. Although we cannot supply full details in the space remaining, we shall end this epilogue with an outline of a proof of the fundamental theorem. We first state it in modern form:

> Any nth degree polynomial ($n \geq 1$) with complex coefficients can be factored into n complex linear factors.

More formally,

> If $P(z) = c_n z^n + c_{n-1} z^{n-1} + \cdots + c_2 z^2 + c_1 z + c_0$, where $n \geq 1$ and $c_n, c_{n-1}, \cdots c_2, c_1, c_0$ are complex numbers with $c_n \neq 0$, then there exist complex numbers $\alpha_1, \alpha_2, \cdots, \alpha_n$ such that $P(z) = c_n(z - \alpha_1)(z - \alpha_2) \cdots (z - \alpha_n)$.

This proposition is seen today in greater generality than in Euler's time. We now transfer it entirely into the realm of complex numbers, where the polynomial with which we begin is no longer required to have *real* coefficients. We thus are considering expressions such as

$$z^7 + 6i\, z^6 - (2 + i)z^2 + 19.$$

In spite of the apparent increase in difficulty, the fundamental theorem still holds. It guarantees that there exist—for this specific example—seven linear factors having, of course, complex coefficients.

To prove the theorem, many prerequisite ideas are necessary. One of the most important is the geometric interpretation of a complex number. The idea—advanced after Euler's death by Caspar Wessel (1745–1818), Jean-Robert Argand (1768–1822), and Gauss—is to introduce the so-called "complex plane" with real axis running horizontally and imaginary axis running vertically. As shown in Figure 6.3, the complex number $z = a + bi$ is represented geometrically as the point (a, b) in the complex plane. Although simple in concept, this link between the algebra and the geometry of complex numbers is an extremely powerful one.

A second prerequisite is the notion of **modulus**, the counterpart of absolute value for real magnitudes. Given a complex number $z = a + bi$, we define the modulus of z by $|z| = \sqrt{a^2 + b^2}$. With the geometric representation above, the modulus is just the distance from the origin to (a, b) and is in this sense the "length" of the complex number.

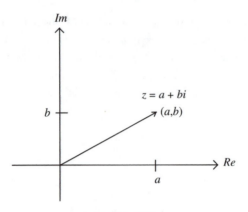

FIGURE 6.3

Meanwhile we introduce the idea of a **complex function** $w = f(z)$, where both the input z and the output w are complex. As we saw in the previous chapter, Euler did just this when he defined the sine, cosine, exponential, and logarithm of complex quantities.

We next specify what it means for a complex function to be **bounded**. Again, our motivation comes from the analogous concept in the reals. We say f is bounded on a set S of complex numbers if there exists a positive real number M so that $|f(z)| \le M$ for all z in S. Geometrically this means that, as z ranges through the complex numbers in S, the corresponding images $f(z)$ fall within a circle of radius M centered at the origin.

Finally, we need the notions of complex limit and complex derivative. The latter, whose obvious ancestor is the derivative from calculus, is defined by

$$f'(z) = \lim_{\Delta z \to 0} \frac{f(z + \Delta z) - f(z)}{\Delta z},$$

provided the limit exists. If the complex derivative exists for all points z in a given set, we say the function f is **analytic** over that set. A function analytic over the set of all complex numbers is said to be **entire**.

With this cursory introduction, we state an important theorem named for the nineteenth-century mathematician Joseph Liouville (1809–1882):

Liouville's Theorem. *An entire, bounded complex function is constant.*

At first, this seems invalid. It says that if a complex function is entire (that is, everywhere differentiable) and if it is also bounded (so that it maps all complex numbers into a bull's-eye of finite radius), then the function must be *constant*—which is to say, trivial.

No one would quibble with the converse: a constant function certainly is entire and bounded. That the implication is reversible comes as a surprise. The skeptic might cite the function $f(x) = \cos x$ as a counterexample. After all, we know from calculus that $\cos x$ is everywhere differentiable (entire) and bounded (because $|\cos x| \leq 1$ for all real numbers x). Yet the cosine function is certainly not a constant.

True enough. But Liouville was considering functions whose domain is the set of all complex numbers. As we saw in the last chapter, Euler successfully defined the cosine function for any complex z . The result was indeed entire and non-constant. But, even as we solved $\cos z = 2$, we could just as well have solved $\cos z = M$ for any $M \geq 1$. Indeed, the cosine function is *not* bounded over the set of all complex numbers and therefore does not provide a counterexample to Liouville's theorem.

We stress that, when done thoroughly, the preliminaries mentioned above would occupy a few months of a complex analysis course. We are condensing shamelessly. Nonetheless, we can now attack the fundamental theorem of algebra. We begin with the key lemma:

Lemma. *If $P(z)$ is a non-constant, complex polynomial, then the equation $P(z) = 0$ has at least one complex solution.*

Proof. We argue by contradiction. If P is never zero, then the reciprocal function $f(z) = 1/P(z)$ is such that:

- f is defined for all complex numbers z ;

- f is entire with derivative, by the chain rule, $f'(z) = -\dfrac{P'(z)}{[P(z)]^2}$;

- f is bounded.

This last condition involves details that would carry us beyond the scope of the chapter.

Therefore f is an entire, bounded function which must be constant by Liouville's theorem. But if f is constant, so is its reciprocal P, a contradiction to the hypothesis that P is a non-constant polynomial.   Q.E.D.

With the lemma behind us, the fundamental theorem of algebra now follows easily:

Theorem. *Any nth degree polynomial ($n \geq 1$) with complex coefficients can be factored into n complex linear factors.*

Proof. Let P be such a polynomial. By the lemma, there exists a complex number α_1 with $P(\alpha_1) = 0$. This means that $z - \alpha_1$ is a factor of $P(z)$, and thus $P(z) = (z - \alpha_1)Q(z)$, where Q is an $(n - 1)$st degree polynomial. We apply the lemma to Q, thereby finding an α_2 so that $P(z) = (z - \alpha_1)(z - \alpha_2)R(z)$, where R has degree $n - 2$. Continuing in this manner, and reducing the degree at each step, we arrive at

$$P(z) = c_n(z - \alpha_1)(z - \alpha_2) \cdots (z - \alpha_n),$$

and the fundamental theorem of algebra is proved.   Q.E.D.

We stress again that filling in all the fine points here is a major enterprise. In particular, Liouville's theorem has a vast mathematical foundation beneath it, a foundation that lay well beyond Euler's eighteenth-century vision.

We have thus concluded these two algebraic stories, tales that can provide a valuable lesson not only about mathematics but also about our mathematical predecessors. In both cases, the final resolution eluded Euler, and the esteemed names in the history books—Abel, Gauss, and Liouville—are of his successors.

That even Euler could fall short may provide a bit of comfort to lesser mathematicians (a category that includes virtually everybody else in history). Yet before we consign his work to the mathematical scrap heap, we might give Euler at least a modest round of applause for the characteristic cleverness, boldness, and mental agility demonstrated in the algebraic results of this chapter. Even when he stumbled, Leonhard Euler put on a great show. Such, perhaps, is a mark of genius.

CHAPTER 7

Euler and Geometry

No one would contend that Euler's mathematical legacy rests primarily upon his contributions to geometry. He lived 20 centuries after the golden age of Greece and decades before the non-Euclidean revolution. In the Geometers' Hall of Fame, where the names of Euclid and Archimedes, Apollonius and Lobachevski are etched large, Leonhard Euler has a niche somewhere off the main corridor.

Still, it is wrong to conclude that Euler ignored this fascinating and timeless subject. On the contrary, four volumes of his *Opera Omnia*, totalling almost 1600 pages, are devoted to geometrical research. Some of his work falls under the heading of "synthetic" geometry—that is, the familiar geometry of Euclid that does not superimpose coordinate axes upon the plane. Most of Euler's geometric papers, however, were of the "analytic" variety in which axes *were* superimposed and in which he freely applied his algebraic powers to treat matters of interest.

In this chapter, we consider two examples of Euler' geometry: his proof of Heron's formula and his discovery of what is now known as the "Euler line" of a triangle. The former was synthetic; the latter was analytic. The former was a new route to a familiar destination; the latter was a new route to an unfamiliar one. Taken together, they suggest that in geometry, as in so many other areas, Leonhard Euler was a force to be reckoned with.

First we must set the stage with a few prerequisites, and this, not surprisingly, necessitates a look backward to the work of the ancient Greeks.

Prologue

The classical period of Greek civilization stretched over many centuries and saw extraordinary advances in science, literature, art, and philosophy. Even today, more than 2000 years later, educated people the world over recognize the names of Homer and Plato and Aristotle. But perhaps no achievement of ancient Greece was more glorious than the creation of demonstrative, axiomatic mathematics. And it is important to recall that for the Greeks, "mathematics" was largely synonymous with "geometry."

Their approach has become the standard: from a carefully selected and limited set of postulates, to deduce ever more sophisticated propositions, with each proof based upon that which has gone before. In this way, the mathematician erects a tower of ideas upon a foundation of simple axioms.

Such a deductive scheme is best seen in Euclid's *Elements*, a work that would influence directly or indirectly all that came after. It was the Islamic mathematician al-Qifti (d. 1248) who observed, "nay, there was no one even of later date who did not walk in [Euclid's] footsteps."[1] These disciples—Archimedes, Apollonius, Ptolemy, and Heron to name a few—would extend the work of Euclid and leave their own indelible marks upon the geometric landscape.

To make understandable the contributions of Euler, we begin with a few results about triangles well known to the Greek geometers. Specifically, we shall review four points—the orthocenter, centroid, circumcenter, and incenter—that exist for any triangle, as well as Heron's theorem for determining a triangle's area from the lengths of its sides.

Special Points of a Triangle

Throughout this chapter, we shall consider a general triangle—$\triangle ABC$—with sides of length a, b, and c, and angles of measure α, β, and γ, as shown in Figure 7.1. (We have drawn a scalene triangle, but proofs can be modified as necessary for right and obtuse triangles as well.) Associated with the triangle are four special points.

1. The **orthocenter** is the intersection of the triangle's three altitudes. It is a standard problem to show that the altitudes indeed meet in a point. The orthocenter is labeled E in Figure 7.2 and throughout this chapter. Incidentally, one can find old textbooks in which the orthocenter is called the "Archimedean point" of the triangle, attesting to its classical provenance.[2]

2. The **centroid** is the intersection of the three medians—i.e., the lines from each vertex to the midpoint of the opposite side. Again, it is a standard

---

[1] T. L. Heath, ed., *The Thirteen Books of Euclid's Elements*, Vol. 1, Dover, New York, 1956, p. 4.
[2] Albert Gminder, *Ebene Geometrie*, R. Oldenbourg, Munich, 1932, p. 294.

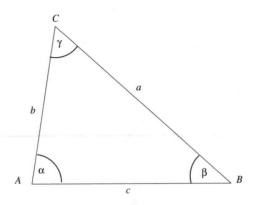

FIGURE 7.1

exercise to show that the medians are concurrent and to prove that the centroid is exactly $\frac{2}{3}$ of the way from each vertex to the opposite midpoint. Labeled F in Figure 7.3, the centroid is also the triangle's center of gravity and thus has an important physical interpretation. Archimedes treated centroids in proposition 13 of the first book of *On the Equilibrium of Planes*, dating from around 225 BCE.[3]

3. The **circumcenter**, as the name suggests, is the center of the triangle's circumscribed circle. It is the intersection of the perpendicular bisectors of the

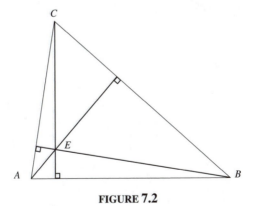

FIGURE 7.2

[3]T. L. Heath, ed., *The Works of Archimedes*, Dover, New York, 1953, pp. 198–201.

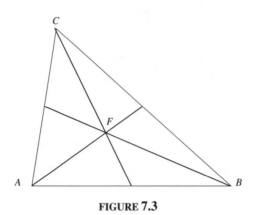

FIGURE 7.3

three sides and is denoted by H in Figure 7.4. Obviously, the radius of the circumscribed circle is $\overline{AH} = \overline{BH} = \overline{CH}$. This matter was discussed by Euclid in proposition 5 of the fourth book of the *Elements*.

4. The **incenter** is the center of the triangle's inscribed circle. As proved in Book IV, proposition 4, of the *Elements*, the incenter is the intersection of the bisectors of the three angles of the triangle. It is denoted by O in Figure 7.5. The radius r of the inscribed circle is the perpendicular distance from O to any of the three sides; that is $r = \overline{OS} = \overline{OT} = \overline{OU}$.

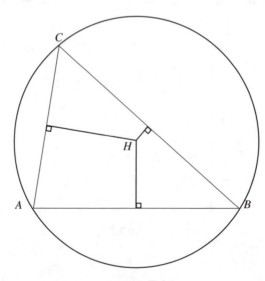

FIGURE 7.4

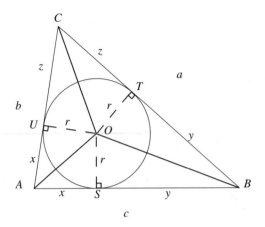

FIGURE 7.5

Of these four points, the incenter is perhaps most special of all. For one thing, it leads to the decomposition of any triangle into a trio of sub-triangles in the following sense (see Figure 7.5):

$$\text{Area}(\triangle ABC) = \text{Area}(\triangle ABO) + \text{Area}(\triangle BOC) + \text{Area}(\triangle AOC)$$

$$= \frac{1}{2}cr + \frac{1}{2}ar + \frac{1}{2}br = r\left(\frac{a+b+c}{2}\right) = rs,$$

where $s = (a+b+c)/2$ is called the **semiperimeter** of $\triangle ABC$. In other words, the area of any triangle is the product of its semiperimeter and its "inradius" (the radius of the inscribed circle). Besides having importance in its own right, this property is critical in proving Heron's formula.

The incenter has further significance. We refer again to Figure 7.5. Because OA not only bisects $\angle BAC$ but is the shared hypotenuse of right triangles $\triangle OSA$ and $\triangle OUA$, their congruence follows at once. We shall let $x = \overline{AS} = \overline{AU}$. Analogous congruence arguments allow us to introduce $y = \overline{BS} = \overline{BT}$ and $z = \overline{CT} = \overline{CU}$.

But there is more. Clearly $a = y + z$, $b = x + z$, and $c = x + y$, and so

$$s = \frac{a+b+c}{2} = \frac{(y+z)+(x+z)+(x+y)}{2} = x + y + z.$$

Consequently,

$$s - a = (x + y + z) - (y + z) = x,$$

$$s - b = (x + y + z) - (x + z) = y,$$

and

$$s - c = (x + y + z) - (x + y) = z.$$

All of this, albeit without the benefit of modern algebraic notation, was known to the Greeks.

Heron's Formula for Triangular Area

Here is one of the treasures of classical geometry. It is clear that the lengths of a triangle's three sides unambiguously determine its area—no surprises there. What *is* surprising, though, is the complexity of the formula necessary to spell this out.

Sometime in the second century, Heron of Alexandria proved that the area of a triangle with sides of length a, b, and c is given by $\sqrt{s(s-a)(s-b)(s-c)}$, where s is the semiperimeter mentioned above. This seems an unreasonably convoluted formula for so basic an idea, but such are the peculiarities of Euclidean geometry.

Heron's proof was wickedly clever. He began by inscribing a circle within his triangle. Constructing a multitude of auxiliary lines, invoking known facts about quadrilaterals inscribed in circles, and making repeated use of similar triangles, he seemed for all the world to be mathematically adrift. Yet he pulled everything together at the last moment to prove his result. Readers are left shaking their heads in amazement at what amounts to a genuine surprise ending.

For three reasons, we shall not consider the specifics of Heron's argument. First, these specifics, although embodying some of the most spectacular reasoning in Greek geometry, would carry us too far afield. Second, the proof is fully discussed elsewhere.[4] Finally, we shall examine Euler's proof—and two others as well—before this chapter has run its course.

Enter Euler

Leonhard Euler, of course, was familiar with Heron's formula, which he called a "memorable rule." In a 1748 paper with the unimaginative title "*Variae demonstrationes geometriae,*" Euler provided a synthetic proof of Heron's formula—more or less in the Greek style—"in which," he pledged, "is seen no

---

[4]See Dunham, *Journey Through Genius*, Chapter 5.

vestige of analysis." As will soon be evident, he delivered on this promise.[5] (Note: we ask forgiveness in adopting the standard practice of Euler's day of not distinguishing between an angle and its measure or between congruence and equality.)

Theorem. *If* $\triangle ABC$ *has sides a, b, and c and semiperimeter* $s = (a+b+c)/2$, *then* Area($\triangle ABC$) $= \sqrt{s(s-a)(s-b)(s-c)}$.

Proof. As always, we begin with $\triangle ABC$ having sides a, b, and c and angles α, β, and γ. Following Heron's lead, Euler first inscribed a circle within the triangle. Let O be the center of the inscribed circle with radius $r = \overline{OS} = \overline{OU}$, as shown in Figure 7.6. Recall from the construction of the incenter that segments OA, OB, and OC bisect the angles of $\triangle ABC$, with $\angle OAB = \alpha/2$, $\angle OBA = \beta/2$, and $\angle OCA = \gamma/2$.

Euler extended BO and constructed a perpendicular from A intersecting this extended line at V. (He drew these internal to the triangle, but the proof may be modified should they fall outside the figure.) Euler denoted by N the

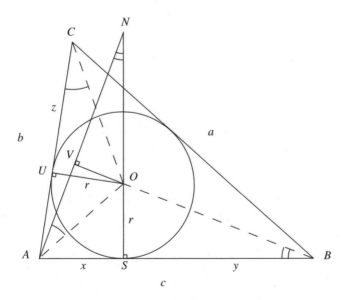

FIGURE 7.6

[5]Euler, *Opera Omnia*, Ser. 1, Vol. 26, pp. 18–22.

intersection of the extensions of segment AV and radius OS. Then he put away his compass and straightedge. These were the only constructions needed in his proof, making the preliminaries far simpler than Heron's from 1600 years before.

Because $\angle AOV$ is an exterior angle of $\triangle AOB$, Euler observed that

$$\angle AOV = \angle OAB + \angle OBA = \alpha/2 + \beta/2.$$

Further, because $\triangle AOV$ is right, he knew that $\angle AOV$ and $\angle OAV$ are complementary. It follows that $\alpha/2 + \beta/2 + \angle OAV = 90°$. But $\alpha/2 + \beta/2 + \gamma/2 = 90°$ as well, so $\angle OAV = \gamma/2 = \angle OCU$.

From this Euler deduced the similarity of right triangles $\triangle OAV$ and $\triangle OCU$ and thus the proportion

$$\overline{AV}/\overline{VO} = \overline{CU}/\overline{OU} = z/r. \tag{7.1}$$

Clearly $\triangle NOV$ and $\triangle NAS$ are similar, as are $\triangle NAS$ and $\triangle BAV$, and so too must be $\triangle NOV$ and $\triangle BAV$. Hence

$$\overline{AV}/\overline{AB} = \overline{OV}/\overline{ON}, \quad \text{or equivalently} \quad \overline{AV}/\overline{OV} = \overline{AB}/\overline{ON}. \tag{7.2}$$

Combining expressions (7.1) and (7.2) yields

$$\frac{z}{r} = \frac{\overline{AB}}{\overline{ON}} = \frac{x+y}{\overline{SN}-r},$$

and so

$$z(\overline{SN}) = r(x + y + z) = rs. \tag{7.3}$$

One last ingredient—the identity of $\overline{SN}$—is needed. Because they are vertical angles, $\angle BOS$ and $\angle VON$ are congruent, and so

$$\angle OBS = 90° - \angle BOS = 90° - \angle VON = \angle ANS.$$

Consequently $\triangle NAS$ and $\triangle BOS$ are similar, from which Euler deduced the proportion $\overline{SN}/\overline{AS} = \overline{BS}/\overline{OS}$. This amounts to $\overline{SN}/x = y/r$, or simply $\overline{SN} = (xy)/r$.

Euler concluded with a flourish:

$$\text{Area}(\triangle ABC) = rs = \sqrt{rs(rs)} = \sqrt{z(\overline{SN})(rs)} \qquad \text{by (7.3)}$$

$$= \sqrt{z\left(\frac{xy}{r}\right)rs} = \sqrt{sxyz} = \sqrt{s(s-a)(s-b)(s-c)}.$$

$$\text{Q.E.D.}$$

This is a most ingenious proof of Heron's theorem. It provides us with a glimpse of Euler the Elegant.

But he was only warming up. In a 1767 paper, Euler again turned his attention to that simplest of plane figures, the triangle. This time, instead of focusing on triangular *area*, he examined relationships among the special points mentioned above. In so doing, he discovered the following remarkable fact: the orthocenter, centroid, and circumcenter of any triangle must lie in a straight line, with the centroid exactly twice as far from the orthocenter as from the circumcenter. This fundamental property of triangles had been overlooked by the thousands of geometers who preceded him, from Euclid to Archimedes to Heron. In recognition of his discovery, the segment containing these three points is now called the triangle's "Euler line."

We shall examine his argument in detail.[6] In the course of the discussion, the reader is urged to observe how Euler used the tools of classical geometry: similar triangles, perpendicular bisectors, and—in a starring role—Heron's formula. But note also that, in contrast to the proof just concluded, Euler also employed the techniques of *analytic* geometry. From the outset, he placed coordinate axes upon the plane and, after some preliminaries, exploited the Cartesian distance formula

$$d = \sqrt{(x_1 - x_2)^2 + (y_1 - y_2)^2}$$

to reach his goal. This gives his work a decidedly algebraic flavor in which the desired conclusion emerges from a blizzard of formulas. In a sense, the proof is a fusion of geometric insight and algebraic perseverance.

Euler began with an arbitrary triangle $\triangle ABC$ having sides of length a, b, and c. Without loss of generality, he placed it in the Cartesian plane with A at the origin and B on the x-axis, as pictured in Figure 7.7.

Euler invoked Heron's formula. For notational ease, we shall let $K = \text{Area}(\triangle ABC)$ so that

$$K = \sqrt{s(s-a)(s-b)(s-c)}$$
$$= \sqrt{\frac{a+b+c}{2} \times \frac{-a+b+c}{2} \times \frac{a-b+c}{2} \times \frac{a+b-c}{2}}.$$

Upon squaring and simplifying, this becomes:

---

[6]Euler, *Opera Omnia*, Ser. 1, Vol. 26, pp. 139–157.

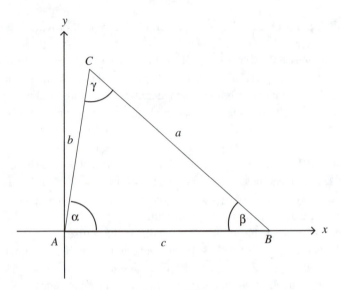

FIGURE 7.7

$$16K^2 = [(b + c) + a][(b + c) - a][a - (b - c)][a + (b - c)]$$
$$= [(b + c)^2 - a^2][a^2 - (b - c)^2]$$
$$= [b^2 + 2bc + c^2 - a^2][a^2 - b^2 + 2bc - c^2] \qquad (7.4)$$
$$= 2a^2b^2 + 2a^2c^2 + 2b^2c^2 - a^4 - b^4 - c^4.$$

This equation would appear repeatedly in Euler's proof.

His strategy was both direct and daunting: he would find the *coordinates* of the three special points in terms of a, b, c, and K and then use these coordinates to determine a relationship among the orthocenter, centroid, and circumcenter.

The Orthocenter (E)

Begin with the orthocenter E in Figure 7.8, where AM and CP are altitudes.
First, apply the Law of Cosines to $\triangle ABC$:

$$a^2 = b^2 + c^2 - 2bc \cos \alpha = b^2 + c^2 - 2bc \left(\frac{\overline{AP}}{b} \right) = b^2 + c^2 - 2c \,\overline{AP}.$$

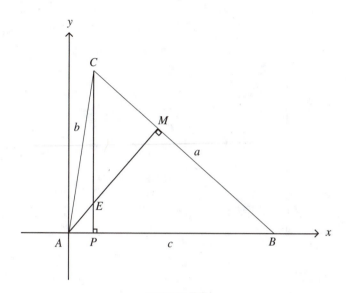

FIGURE 7.8

Thus $\overline{AP} = (b^2 + c^2 - a^2)/2c$. The same argument shows that $\overline{BM} = (a^2 + c^2 - b^2)/2a$. Further,

$$K = \text{Area}(\triangle ABC) = \frac{1}{2}(\overline{BC})(\overline{AM}), \quad \text{so that} \quad \overline{AM} = \frac{2K}{a}.$$

The similarity of $\triangle ABM$ and $\triangle AEP$ implies $\overline{BM}/\overline{AM} = \overline{EP}/\overline{AP}$, and so

$$\overline{EP} = (\overline{BM})(\overline{AP})/\overline{AM} = \left(\frac{a^2 + c^2 - b^2}{2a}\right)\left(\frac{b^2 + c^2 - a^2}{2c}\right) \Big/ \frac{2K}{a}$$

$$= \frac{2a^2b^2 - a^4 - b^4 + c^4}{8cK}$$

$$= \frac{16K^2 - 2a^2c^2 - 2b^2c^2 + 2c^4}{8cK} \quad \text{from (7.4) above}$$

$$= \frac{2K}{c} + \frac{c(c^2 - a^2 - b^2)}{4K}.$$

Hence the orthocenter E has coordinates

$$(\overline{AP}, \overline{EP}) = \left(\frac{b^2 + c^2 - a^2}{2c}, \frac{2K}{c} + \frac{c(c^2 - a^2 - b^2)}{4K}\right).$$

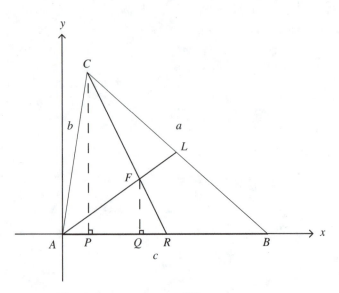

FIGURE 7.9

The Centroid (*F*)

In Figure 7.9, *R* is the midpoint of *AB*, and *L* is the midpoint of *BC*. We have drawn medians *CR* and *AL* meeting at centroid *F*, as well as altitude *CP* which was considered above.

We observe that $K = \text{Area}(\triangle ABC) = \frac{1}{2}(\overline{AB})(\overline{CP})$, so that $\overline{CP} = 2K/c$.

Construct $FQ \perp AB$ and note that $\triangle RQF$ is similar to $\triangle RPC$. It follows that $\overline{RQ}/\overline{RP} = \overline{RF}/\overline{RC} = \frac{1}{3}$ by the well-known theorem about centroids mentioned in the chapter's prologue. Using this and the definition of median, we conclude:

$$\overline{AQ} = \overline{AR} - \overline{RQ} = \frac{1}{2}(\overline{AB}) - \frac{1}{3}(\overline{RP}) = \frac{1}{2}c - \frac{1}{3}(\overline{AR} - \overline{AP})$$

$$= \frac{1}{2}c - \frac{1}{3}\left(\frac{1}{2}c - \frac{b^2 + c^2 - a^2}{2c}\right) \qquad \text{from our discussion of } \overline{AP} \text{ above}$$

$$= \frac{3c^2 + b^2 - a^2}{6c}.$$

This is the abscissa of the centroid.

To determine its ordinate, return to similar triangles $\triangle RQF$ and $\triangle RPC$. From

$$\frac{\overline{FQ}}{\overline{CP}} = \frac{\overline{RF}}{\overline{RC}} = \frac{1}{3}, \quad \text{it follows that} \quad \overline{FQ} = \frac{1}{3}(\overline{CP}) = \frac{2K}{3c}.$$

Thus the coordinates of the centroid are

$$(\overline{AQ}, \overline{FQ}) = \left(\frac{3c^2 + b^2 - a^2}{6c}, \frac{2K}{3c} \right).$$

The Circumcenter (*H*)

The pertinent diagram is now Figure 7.10, where (again) R is the midpoint of AB and D is the midpoint of AC. Through these two points, we construct perpendicular bisectors that meet at the circumcenter H. We have also drawn altitude AM whose length, as derived above, is $\overline{AM} = 2K/a$.

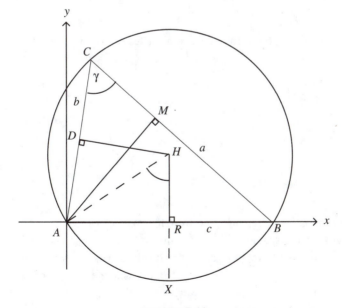

FIGURE 7.10

Applying the Law of Cosines to $\triangle ABC$ yields

$$c^2 = a^2 + b^2 - 2ab \cos \gamma = a^2 + b^2 - 2ab \left(\frac{\overline{CM}}{b} \right)$$

$$= a^2 + b^2 - 2a\,\overline{CM}, \text{ so that}$$

$$\overline{CM} = \frac{a^2 + b^2 - c^2}{2a}.$$

Next, referring to the circumscribed circle, we see that the measure of $\angle ACB$ is half that of the intercepted arc AB—which is to say, the arc AX. But the measure of central angle $\angle AHR$ is that of arc AX as well.

We conclude that $\angle ACB = \angle AHR$ and therefore that $\triangle ACM$ is similar to $\triangle AHR$. From this follows the proportion $\overline{HR}/\overline{AR} = \overline{CM}/\overline{AM}$. Consequently,

$$\overline{HR} = \left(\frac{1}{2}c \right) \left(\frac{a^2 + b^2 - c^2}{2a} \right) \Big/ \frac{2K}{a} = \frac{c(a^2 + b^2 - c^2)}{8K},$$

and so circumcenter H has coordinates

$$(\overline{AR}, \overline{HR}) = \left(\frac{c}{2}, \frac{c(a^2 + b^2 - c^2)}{8K} \right).$$

To summarize, Euler found the coordinates of the three key points to be

orthocenter E : $\left(\dfrac{b^2 + c^2 - a^2}{2c}, \dfrac{2K}{c} + \dfrac{c(c^2 - a^2 - b^2)}{4K} \right)$

centroid F : $\left(\dfrac{3c^2 + b^2 - a^2}{6c}, \dfrac{2K}{3c} \right)$

circumcenter H : $\left(\dfrac{c}{2}, \dfrac{c(a^2 + b^2 - c^2)}{8K} \right).$

Unfortunately he remained far from his goal, for he still had to determine the lengths of the segments $\overline{EF}$, $\overline{EH}$, and $\overline{FH}$. A lesser mathematician may have despaired, but Euler proceeded merrily along. His only concession to the tedium—and a minor one at that—was to work with the *squares* of the lengths rather than the lengths themselves. Referring to the coordinates above, we take a deep breath and follow him on his journey.

$$(\overline{EF})^2 = \left[\frac{b^2 + c^2 - a^2}{2c} - \frac{3c^2 + b^2 - a^2}{6c}\right]^2$$

$$+ \left[\frac{2K}{c} + \frac{c(c^2 - a^2 - b^2)}{4K} - \frac{2K}{3c}\right]^2$$

$$= \left[\frac{b^2 - a^2}{3c}\right]^2 + \left[\frac{4K}{3c} + \frac{c(c^2 - a^2 - b^2)}{4K}\right]^2$$

$$= \frac{(b^2 - a^2)^2 + 16K^2}{9c^2} + \frac{2c^2 - 2a^2 - 2b^2}{3}$$

$$+ \frac{c^2(c^4 + a^4 + b^4 - 2a^2c^2 - 2b^2c^2 + 2a^2b^2)}{16K^2}.$$

Heron's formula—as expressed in equation (7.4)—can be used to simplify the right-hand numerator, giving

$$(\overline{EF})^2 = \frac{(b^2 - a^2)^2 + 16K^2}{9c^2} + \frac{2c^2 - 2a^2 - 2b^2}{3} + \frac{c^2(4a^2b^2 - 16K^2)}{16K^2}$$

$$= \frac{(b^2 - a^2)^2 + 16K^2}{9c^2} - \frac{2a^2 + 2b^2 + c^2}{3} + \frac{a^2b^2c^2}{4K^2},$$

which (thankfully) is as far as we shall need to take it.

Next up was:

$$(\overline{EH})^2 = \left[\frac{b^2 + c^2 - a^2}{2c} - \frac{c}{2}\right]^2$$

$$+ \left[\frac{2K}{c} + \frac{c(c^2 - a^2 - b^2)}{4K} - \frac{c(a^2 + b^2 - c^2)}{8K}\right]^2$$

$$= \left[\frac{b^2 - a^2}{2c}\right]^2 + \left[\frac{2K}{c} + \frac{3c(c^2 - a^2 - b^2)}{8K}\right]^2$$

$$= \frac{(b^2 - a^2)^2 + 16K^2}{4c^2} + \frac{3c^2 - 3a^2 - 3b^2}{2}$$

$$+ \frac{9c^2(c^4 + a^4 + b^4 - 2a^2c^2 - 2b^2c^2 + 2a^2b^2)}{64K^2}.$$

As before, we apply (7.4) to simplify the right-hand numerator:

$$(\overline{EH})^2 = \frac{(b^2 - a^2)^2 + 16K^2}{4c^2} + \frac{3c^2 - 3a^2 - 3b^2}{2} + \frac{9c^2(4a^2b^2 - 16K^2)}{64K^2}$$

$$= \frac{(b^2 - a^2)^2 + 16K^2}{4c^2} - \frac{6a^2 + 6b^2 + 3c^2}{4} + \frac{9a^2b^2c^2}{16K^2}.$$

Lastly, we determine:

$$(\overline{FH})^2 = \left[\frac{3c^2 + b^2 - a^2}{6c} - \frac{c}{2}\right]^2 + \left[\frac{2K}{3c} - \frac{c(a^2 + b^2 - c^2)}{8K}\right]^2$$

$$= \left[\frac{b^2 - a^2}{6c}\right]^2 + \frac{4K^2}{9c^2} - \frac{a^2 + b^2 - c^2}{6}$$

$$+ \frac{c^2(a^4 + b^4 + c^4 + 2a^2b^2 - 2a^2c^2 - 2b^2c^2)}{64K^2}$$

$$= \frac{(b^2 - a^2)^2 + 16K^2}{36c^2} - \frac{a^2 + b^2 - c^2}{6} + \frac{c^2(4a^2b^2 - 16K^2)}{64K^2} \quad \text{by (7.4)}$$

$$= \frac{(b^2 - a^2)^2 + 16K^2}{36c^2} - \frac{2a^2 + 2b^2 + c^2}{12} + \frac{a^2b^2c^2}{16K^2}.$$

And now Euler pulled together these algebraic/geometric preliminaries to reach his conclusion.

Theorem. *In any triangle, the orthocenter (E), the centroid (F), and the circumcenter (H) are collinear, with $\overline{EF} = 2(\overline{FH})$ and $\overline{EH} = 3(\overline{FH})$.*

Proof. We let $d = \overline{FH}$ and consider the results above:

$$(\overline{EF})^2 = \frac{(b^2 - a^2)^2 + 16K^2}{9c^2} - \frac{2a^2 + 2b^2 + c^2}{3} + \frac{a^2b^2c^2}{4K^2}$$

$$= 4\left[\frac{(b^2 - a^2)^2 + 16K^2}{36c^2} - \frac{2a^2 + 2b^2 + c^2}{12} + \frac{a^2b^2c^2}{16K^2}\right] = 4(\overline{FH})^2.$$

Therefore $\overline{EF} = 2(\overline{FH}) = 2d$. This means that the centroid is twice as far from the orthocenter as it is from the circumcenter.

In addition,

$$(\overline{EH})^2 = \frac{(b^2 - a^2)^2 + 16K^2}{4c^2} - \frac{6a^2 + 6b^2 + 3c^2}{4} + \frac{9a^2b^2c^2}{16K^2}$$

$$= 9\left[\frac{(b^2 - a^2)^2 + 16K^2}{36c^2} - \frac{2a^2 + 2b^2 + c^2}{12} + \frac{a^2b^2c^2}{16K^2}\right] = 9(\overline{FH})^2,$$

and so $\overline{EH} = 3(\overline{FH}) = 3d$.

These calculations show that the three points are different unless $d = 0$, a phenomenon that occurs only for an equilateral triangle. More significantly, they guarantee that E, F, and H fall on the same line, for

$$\overline{EH} = 3d = 2d + d = \overline{EF} + \overline{FH},$$

as shown in Figure 7.11. Were the points not collinear, this would contradict (appropriately enough) the triangle inequality.   Q.E.D.

Here, then, is the origin of the "Euler line," a property of triangles that David Wells rightly called Euler's "celebrated theorem."[7] It is for this, as much as anything, that Euler has his niche in the Geometer's Hall of Fame.

We have included this argument alongside Euler's proof of Heron's formula for two reasons. First, they demonstrate that the great eighteenth-century analyst, algebraist, and number theorist was capable of doing significant geometry. His mathematical versatility knew no bounds.

But these two proofs serve an additional purpose: to represent the opposing sides of a controversy dating back to the time of Descartes. The issue under debate was the role to be played in geometry by that Johnny-come-lately, algebra.

Consider Euler's proof of Heron's formula. This was an example of (in Euler's words) "pure" geometry. The Euler line, by contrast, was a creature of the Cartesian revolution. It would have been all Greek to Euclid.

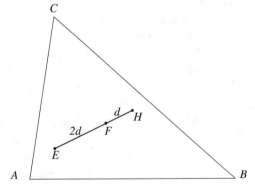

FIGURE 7.11

[7]David Wells, *The Penguin Book of Curious and Interesting Geometry*, Penguin, New York, 1991, p. 69.

In the opinion of certain mathematicians of the past, the latter was inferior to the former. Their objection to analytic geometry was largely an aesthetic one, for pure geometry often requires a leap of insight that goes by the name "inspiration." How, for instance, did Euler know to construct ΔNOV in his proof of Heron's formula? How did he know which similar triangles would prove fruitful in leading him to the desired conclusion? How, in short, did he know what to do?

Ultimately, the answer to this question lies in the mysterious realm of the human imagination. One might just as well ask of Shakespeare why he put the balcony scene in *Romeo and Juliet*. Yes, it was an important element of plot and character development, but the lovers could have met in a garden or forest or piazza. The choice of *balcony* was an aesthetic one needing no justification other than it worked—and worked beautifully. We are left to appreciate the sheer artistry of the moment. In just this way, we appreciate the aesthetic elements present in the best of synthetic geometry.

Contrast this with the second argument above. Having determined that he needed the coordinates of the three special points, Euler ground them out algebraically.[8] It was (we shall be candid here) brutal. His calculations, in an apt description by Eduard Study, resounded with "the clatter of the coordinate mill." Indeed, one could ask whether analytic geometry was really *geometry* at all. Lacking grace and elegance, dependent upon what Carnot called "the hieroglyphics of analysis," was it not merely an application of unrelenting algebraic force?[9]

There was a time when such contempt for analytic geometry was intense, prompting mathematicians like Michel Chasles (1793–1880), Gaspard Monge (1746–1818), and Jakob Steiner (1796–1863) to reject such methods as being ugly, if not unsporting. Just as we would condemn a mountaineer who reached the summit of Everest by parachuting from a passing airplane, so too did purists scorn geometric proofs that were as algebraic as the discovery of the Euler line. Some even seemed to regret the invention of analytic geometry and, in the words of historian Morris Kline, sought "revenge" upon Descartes.

Not surprisingly, there was a counterargument. Analytic geometry, after all, had unquestioned power, amply illustrated in the proof above. It provided a general method. It established connections whose geometric significance may

---

[8] For a much shorter derivation of the Euler Line, see J. Ferrer, "A Vector Approach to Euler's Line of a Triangle," *The American Mathematical Monthly*, Vol. 99, No. 7, 1992, pp. 663–664.
[9] Kline, p. 835.

have been unclear but whose algebraic validity was indisputable. And it did not rely upon divinely inspired bursts of insight. Jean-Victor Poncelet (1788–1867), no great fan of coordinate geometry, conceded this latter point when he wrote:

> While analytic geometry offers by its characteristic method general and uniform means of proceeding to the solution of questions..., the other [classical geometry] proceeds by chance; its way depends completely on the sagacity of those who employ it.[10]

Time has a way of turning heated controversies into harmless footnotes. Nowadays few mathematicians feel compelled to take sides in the battle between synthetic and analytic geometry. Quite the contrary, the two Eulerian proofs we examined perfectly illustrate the value of having multiple weapons in one's geometric arsenal. If in the first proof we encountered Euler the Elegant, in the second we most certainly met Euler the Toiler.

Epilogue

As promised, the epilogue will provide two additional proofs of Heron's formula. Both begin—as did Heron and Euler—at the triangle's incenter. But now we shall exploit the power of *trigonometry* and thereby streamline the proofs significantly.[11]

Theorem. $Area(\triangle ABC) = \sqrt{s(s-a)(s-b)(s-c)}$.

Proof. In Figure 7.12 we consider $\triangle ABC$ in two different ways, both of which we have seen before. On the left we have drawn the triangle with its incenter O; on the right we have included the altitude CP of length h.

From the left-hand diagram, it is clear that

$$\sin\frac{\alpha}{2} = \frac{r}{\sqrt{r^2 + x^2}} \text{ and } \cos\frac{\alpha}{2} = \frac{x}{\sqrt{r^2 + x^2}},$$

and from the right-hand figure we see that $\sin\alpha = h/b$. Therefore,

---

[10]Ibid., p. 834.
[11]William Dunham, "An Ancient/Modern Proof of Heron's Formula," *Mathematics Teacher*, Vol. 78, No. 4, 1985, pp. 258–259.

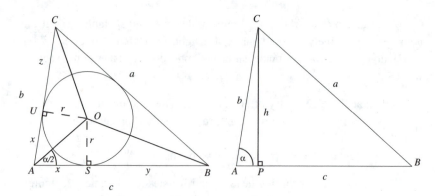

FIGURE 7.12

$$h = b \sin \alpha = b \left[2 \sin \frac{\alpha}{2} \cos \frac{\alpha}{2} \right] = b \frac{2rx}{r^2 + x^2}$$

$$= (x + z) \frac{2rx}{r^2 + x^2},$$

and so

$$rs = \text{Area}(\triangle ABC) = \frac{1}{2}(\overline{AB})(\overline{CP}) = \frac{1}{2}(x + y)h = \frac{1}{2}(x + y)(x + z) \frac{2rx}{r^2 + x^2}.$$

Cross-multiply and simplify to get:

$$s(r^2 + x^2) = x(x + y)(x + z) = x[x(x + y + z) + yz] = x[xs + yz].$$

It follows that $sr^2 + sx^2 = sx^2 + xyz$, or equivalently $sr^2 = xyz$. But then

$$\text{Area}(\triangle ABC) = rs = \sqrt{s(sr^2)} = \sqrt{s(xyz)} = \sqrt{s(s - a)(s - b)(s - c)},$$

which is Heron's formula.   Q.E.D.

That proof shows the value of combining the old (Heron's inscribed circle) and the new (trigonometry). But for sheer efficiency, it is hard to beat the following:[12]

Theorem. $Area(\triangle ABC) = \sqrt{s(s - a)(s - b)(s - c)}.$

[12] Barney Oliver, "Heron's Remarkable Triangular Area Formula," *Mathematics Teacher*, Vol. 86, No. 2, pp. 161–163.

Proof. Note that if δ and θ are positive quantities such that $\delta + \theta = \pi/2$, then $(\tan \delta)(\tan \theta) = (\tan \delta)(\cot \delta) = 1$. So, if $\triangle ABC$ has angles α, β, and γ, we know

$$1 = \tan\left(\frac{\alpha}{2}\right) \tan\left(\frac{\beta}{2} + \frac{\gamma}{2}\right) = \tan\left(\frac{\alpha}{2}\right) \frac{\tan \beta/2 + \tan \gamma/2}{1 - (\tan \beta/2)(\tan \gamma/2)},$$

and thus

$$\tan\left(\frac{\alpha}{2}\right) \tan\left(\frac{\beta}{2}\right) + \tan\left(\frac{\alpha}{2}\right) \tan\left(\frac{\gamma}{2}\right) + \tan\left(\frac{\beta}{2}\right) \tan\left(\frac{\gamma}{2}\right) = 1.$$

Apply this identity to our triangle with its inscribed circle as seen in Figure 7.13:

$$\frac{r}{x} \times \frac{r}{y} + \frac{r}{x} \times \frac{r}{z} + \frac{r}{y} \times \frac{r}{z} = 1.$$

This simplifies to $xyz = r^2(x + y + z) = r^2 s$, and the proof follows from here as in the previous theorem. Q.E.D.

Having proved Heron's formula three times in this chapter, we trust that the reader wholeheartedly believes it.

And what of the Euler line? Recall that Euler published his discovery in 1767. It is interesting to note that, in the century that followed, geometry experienced a kind of renaissance. What had seemed like a dead subject was suddenly revitalized. Surely it would be incorrect to attribute this entirely to

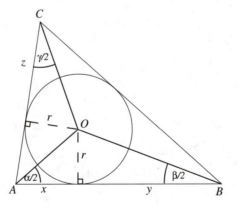

FIGURE 7.13

Euler's influence, yet his successors could not fail to have noticed that this dominant mathematician had found plane geometry so worthy of his attention.

Of course, much of the renewed interest can be explained by the discovery of non-Euclidean geometry in the first half of the nineteenth century and by the nearly simultaneous flowering of projective geometry. But a great deal of scholarship was directed toward good old Euclidean geometry. "It is truly astonishing," wrote historian of mathematics Florian Cajori, "that ... new theorems should have been found relating to such simple figures as the triangle and circle, which had been subject to such close examination by the Greeks and the long line of geometers which followed."[13]

Within a century of the Euler line, mathematicians had uncovered curious new properties of the triangle, and Euclidean geometry now featured terms like the Nagel point, the Georgonne point, and the Feuerbach circle. Even the French emperor got into the act with a result today known as "Napoleon's Theorem." If the Greeks gave us the Golden Age of geometry, then the century after Euler may well be regarded as a Silver Age.

Here we shall discuss only one such topic, the Feuerbach circle, for it relates directly to Euler's line. We begin with $\triangle ABC$ in Figure 7.14. Consistent with our notation above, we let R, L, and D be the midpoints of the three sides and AM, BY, and CP be the three altitudes intersecting at orthocenter E. We

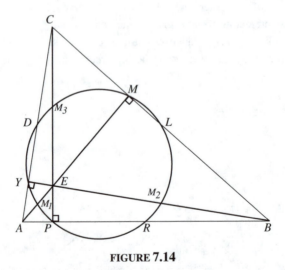

FIGURE 7.14

[13] Florian Cajori, *A History of Mathematics*, Macmillan, New York, 1922, p. 297.

bisect the segments extending from each vertex to the orthocenter, letting M_1 be the midpoint of AE; M_2 be the midpoint of BE; and M_3 be the midpoint of CE.

Now we are in a position to state the theorem in the words of its discoverers, Poncelet and C. J. Brianchon (1785–1864):

> The circle which passes through the feet of the perpendiculars dropped from the vertices of any triangle on the sides opposite them, passes also through the midpoints of these sides as well as through the midpoints of the segments which join the vertices to the point of intersection of the perpendiculars.[14]

That is, the nine points M, Y, and P (the feet of the three altitudes) ; R, L, and D (the midpoints of the sides); and M_1, M_2, and M_3 all lie on a single circle. Furthermore—and here the plot thickens—the center of this circle is the midpoint of the Euler line, and its radius is half the radius of the circumscribed circle. Within the apparently simple triangle, geometers had unearthed a wonderful tangle of relationships.

Unfortunately the Feuerbach circle, like many things in mathematics, is misnamed. As noted, it was first described in 1821 by Poncelet and Brianchon. A year later Karl Wilhelm Feuerbach (1800–1834) stumbled upon related, although not quite identical, ideas. Inspired by Euler's work, Feuerbach published a paper and somehow got *his* name attached to the Poncelet/Brianchon circle.

Today, one is apt to see this called "the nine-point circle," a less colorful if more accurate appellation. Call it what you will, its existence remains intriguing. After all, these nine points were perfectly well understood by the Greeks, yet no one noticed the circle containing them until 1821. This again reminds us of one of the eternal charms of mathematics: its ability to surprise.

If the nine-point circle seems strange, a later discovery must rank among the most peculiar theorems in all of geometry. The result, known as Morley's theorem, was announced in 1899 by American mathematician Frank Morley (1860–1937). It warrants a quick digression before this chapter concludes.

Whereas Euclid found a triangle's incenter by bisecting the three angles, Morley asked what would happen if we began by *trisecting* them. Of course, by his time the impossibility of a compass and straightedge trisection had been proved, but the trisecting lines surely *exist* even if they cannot be constructed with Euclidean tools. Morley asked what would happen when these trisectors meet one another within the triangle (see Figure 7.15).

---

[14]Smith, p. 337.

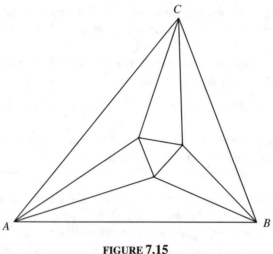

FIGURE 7.15

The answer is spectacular: regardless of the nature of the original triangle, the three pairwise intersections of trisectors always form the vertices of an equilateral triangle! Distort the original and this internal equilateral triangle may change orientation or size, but those three equal sides and three 60° angles are always present. For some non-intuitive reason, the trisectors create an invariant regularity within.[15] Morley's theorem is startling, difficult to prove, and utterly beautiful.

Geometry has been described as the branch of mathematics "most subject to changing tastes from age to age."[16] From its pinnacle in classical times to its renaissance in the nineteenth century to its relatively neglected state at present, geometry has certainly seen wide swings of popularity. But as the theorems of this chapter make clear, Euclid's geometry runs deeper than one might at first imagine, concealing strange and wonderful properties behind misleadingly "elementary" façades. The beauty of geometry well-done—from Heron to Euler, from Poncelet to Morley—is indisputable.

Those who remain skeptical are advised to draw a large triangle, get out compass and straightedge, and watch the orthocenter, centroid, and circumcenter line up as if by magic amid the swirl of necessary arcs. The Euler line remains a wonder of geometry and a fitting tribute to its discoverer, an individual for whom variety was indeed the spice of mathematics.

---

[15] Kline, pp. 839–840.
[16] Carl Boyer & Uta Merzbach, *A History of Mathematics*, 2nd ed., Wiley, New York, 1991, p. 533.

CHAPTER 8

Euler and Combinatorics

Combinatorics is an important and sprawling branch of discrete mathematics, one of whose primary objectives is to count finite collections of items. This may seem an easy task; counting, after all, does not require higher-order thinking skills.

If, for example, three customers enter a bakery, we may ask in how many different ways they can queue up at the counter. Labeling the customers a, b, and c, we list all possible line-ups as

$$abc, \quad acb, \quad bac, \quad bca, \quad cab, \quad cba$$

and count these to see that there are six possibilities for the bakery line. Easy as pie.

Far less elementary is the following question: If the same bakery carries 15 kinds of doughnuts, in how many ways can a hungry customer select a dozen of these? Here we must consider everything from a purchase where all the doughnuts are alike to one where they are all of different varieties.

This time there are 9,657,700 possibilities. Put another way, there are almost ten million distinct boxes of doughnuts a customer can carry out of the bakery. This answer obviously was *not* obtained by listing all options and counting them as we did in the first problem. Rather, the doughnut question required combinatorial theory.

Although combinatorics emerged as a formal branch of mathematics fairly recently, problems of enumeration have a long history. It should come as no surprise that Leonhard Euler made his share of contributions to the subject. After a brief discussion of the work of his predecessors, we shall consider in detail two fascinating investigations by Euler: one involving what he called a "curious" problem of restricted permutations, and the other featuring his insightful analysis of number partitions.

Prologue

It is impossible to specify the earliest combinatorial discovery, but problems of this kind appeared in the work of the Indian mathematician Bhaskara (1114–ca 1185). Some time later, in southern France, Levi ben Gerson (1288–1344) wrote the *Maasei Hoshev* or *The Art of the Calculator*, the greatest pre-modern treatise on the subject.[1] There, ben Gerson proved some of the key formulas of elementary combinatorics while using an early version of mathematical induction.

Over the following centuries, other mathematicians dabbled in this subject. A generation before Euler, basic combinatorial results were assembled by Jakob Bernoulli in his classic *Ars Conjectandi*, written in the late 1600s but published posthumously in 1713. As the title suggests, the book was devoted to the theory of probability, and its most spectacular result, now known as the Law of Large Numbers, certainly stands as a pillar of that theory. In the course of calculating probabilities, Bernoulli needed to count *possibilities*—that is, to determine the number of possible arrangements or selections of certain items. This he discussed in "The Doctrine of Permutations and Combinations," the second chapter of *Ars Conjectandi*, where he gave a valuable overview of combinatorics from around 1700.

Bernoulli began by acknowledging the "infinite variety" of Nature which overwhelms the human capacity to understand, let alone explicitly list, all possibilities for a given situation. However, "most useful in the service of counting is the art, called *Combinatoria*, which remedies this defect of our minds."[2]

In the interest of remedying defects, Bernoulli first took up **permutations**, that is, selections of items in which the order of arrangement matters. He provided the following rule: n distinct items can be lined up in order in $n(n - 1)(n - 2) \cdots 3 \cdot 2 \cdot 1$ different ways. This expression, of course, is now written more concisely as $n!$. In order to help the computationally challenged, Jakob attached a table of values of $n!$ up to $12! = 479{,}001{,}600$.

This permutation rule is an immediate consequence of the multiplication principle of combinatorics, surely one of the most innocent-*looking* yet least innocent theorems in all of mathematics. It says that if we have a two-step process, the first step of which can be done in m different ways and the second

---

[1] See Katz, p. 214 and pp. 278–282.
[2] Jakob Bernoulli, *Ars Conjectandi*, p. 73.

step of which can be done in n different ways, then the entire process can be completed in $m \times n$ different ways.

For instance, if we have three upper-case letters—A, B, and C—and four lower-case letters—e, f, g, and h—the number of ways of choosing first an upper- and then a lower-case letter is $3 \times 4 = 12$, as is readily confirmed by listing the possibilities:

$$
\begin{array}{ccc}
A\,e & B\,e & C\,e \\
A\,f & B\,f & C\,f \\
A\,g & B\,g & C\,g \\
A\,h & B\,h & C\,h
\end{array}
$$

Note that the three upper-case letters determine the three columns and the four lower-case letters generate the four rows, giving us a 3-by-4 array with $3 \times 4 = 12$ entries. The multiplication principle extends in the obvious manner to three-step, four-step, or any multi-step process.

To illustrate the principle, Bernoulli asked how many ways one can play some or all notes from an octave of a "pneumatic organ."[3] For each of the 12 notes from one C to the next, there are two choices: either play it or not. Thus, there are $2 \times 2 \times \cdots \times 2 = 2^{12} = 4096$ ways to proceed. However, because one of these amounts to playing *none* of the notes—and thus sitting in complete silence—it must be discarded, leaving 4095 possible sounds. Bernoulli failed to mention that most of these would be unendurably cacophonous.

Returning to permutations of n distinguishable items, Bernoulli observed that there are n ways to choose the first item, then $n-1$ ways to pick the next item, then $n-2$ ways to pick the next, and so on. The multiplication principle guarantees that there are $n(n-1)(n-2)\cdots 3 \cdot 2 \cdot 1 = n!$ permutations altogether. If we choose only $r \leq n$ items, similar reasoning shows that there are $n(n-1)(n-2)\cdots(n-r+1)$ permutations of length r.

Jakob Bernoulli also considered **combinations**—that is, unordered selections or *subsets*—chosen from a larger assemblage. Unlike the line at a bakery, a combination of people makes no distinction regarding order; what matters is who is present, not where they stand.

In modern notation, we let $\binom{n}{r}$ denote the number of subsets of r items chosen from among n distinguishable items. To determine a formula for this number, we exploit the multiplication principle as follows:

---

[3]Ibid., p. 85.

We shall count, in two separate ways, the number of *permutations* of r items chosen from these n items. On the one hand, we just observed that there are $n(n-1)\cdots(n-r+1)$ such ordered arrangements. On the other hand, we can regard the creation of an ordered arrangement as a two-step process: first we grab a handful of r items without regard to order; and, second, we arrange these r items in a specific manner. In our notation, there are $\binom{n}{r}$ subsets of r items and each of these can be arranged in $r!$ different ways. By the multiplication principle, there are $\binom{n}{r} \times r!$ permutations altogether.

Having counted the permutations twice, we now equate the results:

$$n(n-1)\cdots(n-r+1) = \binom{n}{r} \times r! \text{ and so } \binom{n}{r} = \frac{n(n-1)\cdots(n-r+1)}{r!}.$$

Thus number of different subsets of four letters of the alphabet is

$$\binom{26}{4} = \frac{26 \times 25 \times 24 \times 23}{4 \times 3 \times 2 \times 1} = 14{,}950.$$

This rule for combinations had been recognized by Bhaskara and proved by Levi ben Gerson long before. Here Jakob Bernoulli was providing background for the beginner.

Pushing deeper into the subject, Bernoulli considered the number of combinations of r items that can be chosen from n items if we allow the n items to be reused as often as we wish. This is exactly the situation facing our customer with a taste for doughnuts. Because the order in which the confections are put into the box obviously does not matter, we have a combination rather than a permutation. However, having chosen a glazed doughnut, the customer may certainly choose another glazed doughnut, and then another. In this sense, an object once chosen remains available to be chosen again.

Under the "doughnut scenario," how many combinations are possible? Jakob Bernoulli presented the following rule:

> Let two increasing arithmetic progressions be formed, the first starting from the number of things to be combined, the other from unity, in both of which the common difference is unity, and let each have as many terms as the degree of the combination has units. Then let the product of the terms of the first progression be divided by the product of the terms of the second progression, and the quotient will be the desired number of combinations.[4]

---

[4]Smith, p. 275.

His meaning here is anything but transparent. Indeed, it is difficult to understand both what he is saying and why it works. We shall begin with the former.

If we are counting combinations of r things chosen from n distinguishable items which may be reused, Bernoulli's two arithmetical progressions are:

$$n \quad n+1 \quad n+2 \quad \cdots \quad n+r-2 \quad n+r-1$$
$$1 \quad 2 \quad 3 \quad \cdots \quad r-1 \quad r \,.$$

To find the number of combinations, he said that we should divide the product of the terms in the first progression by the product of those in the second, to get

$$\frac{n \times (n+1) \times (n+2) \times \cdots \times (n+r-2) \times (n+r-1)}{1 \times 2 \times \cdots \times (r-1) \times r}.$$

A glance above will convince the reader that, in modern notation, this is $\binom{n+r-1}{r}$.

But is it correct? And if so, why?

Perhaps the easiest way to verify Bernoulli's rule is to begin with a simple case. Suppose we intend to buy three doughnuts from a bakery that has four different varieties: glazed, sugared, chocolate, and plain—denoted g, s, c, and p. Of course, if there were a single doughnut of each type, we would have $\binom{4}{3} = 4$ possible ways to make our purchase.

Instead, we postulate that there is an inexhaustible supply of each variety (as one expects from a well-stocked bakery). For the sake of explanation, suppose the baker does not put the doughnuts directly into the box but first puts them into individual bags: one for glazed, one for sugared, one for chocolate, and one for plain. When the customer makes a selection, the doughnuts are placed into the proper bags and then the four bags—empty or not—are placed into the box.

If, for instance, we ordered two glazed and a sugared, we would depict the box as

$$g \quad g \,|\, s \,|\, \,|\,.$$

Here a vertical bar (|) represents the division between bags, the last two of which are empty. In fact, once we agree that the bags are presented from left-to-right as glazed, sugared, chocolate, and plain, this could be further simplified to X X | X | | , meaning two doughnuts in the first bag, one in the second, and the last two empty. Likewise, a selection of one glazed, one chocolate, and one plain would be depicted as X | | X | X, and a choice of three chocolate would be | | X X X | .

A moment's thought reveals that under this scheme we have only to fill a string of six slots (i.e., _ _ _ _ _ _) with three Xs and three vertical bars. Each different assignment of three Xs and three bars to these slots represents a different purchase of doughnuts and, conversely, each purchase corresponds to precisely one such arrangement.

So, the original question is transformed into the more abstract one of counting the number of ways to fill six positions with three Xs and three bars. Of course, once we select the three slots to be filled with an X, there is nothing more to do, for the other three will contain a bar.

We conclude that there are $\binom{6}{3} = \frac{6 \times 5 \times 4}{3 \times 2 \times 1} = 20$ different ways to locate the three Xs and, as a consequence, 20 ways to buy the doughnuts. For the skeptic, here they are:

$$g\,g\,g \quad g\,g\,s \quad g\,g\,c \quad g\,g\,p \quad g\,s\,s \quad g\,c\,c \quad g\,p\,p \quad g\,s\,c \quad g\,s\,p \quad g\,c\,p$$
$$s\,s\,s \quad s\,s\,c \quad s\,s\,p \quad s\,c\,c \quad s\,p\,p \quad s\,c\,p \quad c\,c\,c \quad c\,c\,p \quad c\,p\,p \quad p\,p\,p$$

Following exactly this line of reasoning, one derives the general rule. We have n items, which may be reused, and we wish to select r of these without regard to the order of selection. Exactly as above, we abstract this to a string of r Xs and $n - 1$ bars, for a total of $n + r - 1$ slots. From these we must select the r positions to be filled with Xs, a choice that can be made in $\binom{n+r-1}{r}$ ways. This is exactly what Jakob Bernoulli had prescribed in the *Ars Conjectandi*.

So, referring to the problem from the beginning of the chapter, we can choose a dozen doughnuts ($r = 12$) from a well-stocked bakery with fifteen varieties ($n = 15$) in

$$\binom{15 + 12 - 1}{12} = \binom{26}{12} = 9{,}657{,}700$$

different ways. As the mouth waters, the mind reels.

It is hoped that this minicourse in combinatorics will convince the reader that the subject is one in which apparently tame situations prove to be anything but. The problem of counting possibilities is not one that requires deep insights into calculus or complex numbers, yet it possesses its own set of challenges every bit as demanding as one finds in these seemingly more "advanced" branches of mathematics.

One who never tired of challenges was Leonhard Euler. His eclectic mathematical interests led him to problems of enumeration, and, as we shall see below, he not only asked some intriguing questions but found some beautiful answers.

Enter Euler

In October of 1779, Euler prepared for the St. Petersburg Academy a paper on "a curious question from the doctrine of combinations."[5] So great was the backlog of Euler's writings that this was not published in the Academy's Memoirs until 1811, more than a quarter-century after his death. This delay notwithstanding, he provided an example not only of a "curious" problem but also of a method of attack—recursion—that has become one of the primary weapons in the combinatorial arsenal.

The problem had been posed, and solved, decades earlier. In 1708, Pierre Remond de Montmort (1678–1719) discussed a card game called *Treize* or, more descriptively, *Rencontre* ("coincidence"). The rules are these: a player shuffles a deck of 13 cards from ace to king, then deals them one at a time, saying "one" with the first card dealt, "two" with the second, and so on. Victory is achieved if at some point the card overturned exactly matches the word spoken—for instance, if the fourth card turned over is a "four" or the eleventh card laid down is a "jack." Should such a coincidence occur, the dealer wins.

Montmort determined the probability of winning, thereby earning for himself a place of honor in the history of combinatorics. Over the next few years, he corresponded on this matter with Nicolaus Bernoulli, and it was independently considered by Abraham De Moivre in his book *The Doctrine of Chances* of 1718. Obviously, the problem of coincidences had been laid to rest long before Euler came along. Why, then, are we including it here?

There are several reasons.

First, it is a problem with a non-trivial, even *dramatic*, conclusion that appealed to Euler's sense of wonder.

Second, Euler apparently had no knowledge of the work of his predecessors. In the words of Anders Hald, chronicler of the early history of combinatorics,

> The problem of coincidences is an example of a problem occurring in many different contexts and therefore solved by many different authors, many of them unaware of the fundamental contributions by Montmort, Nicholas Bernoulli, and de Moivre.[6]

Euler, in his discussion of the problem, makes no mention of these earlier pioneers. Given that he was invariably generous in bestowing praise and sharing

---

[5] Euler, *Opera Omnia*, Ser. 1, Vol. 7, pp. 435–440.

[6] Anders Hald, *A History of Probability and Statistics and their Applications before 1750*, Wiley, New York, 1990, p. 340.

credit, Euler's omission suggests that he did not know he had been scooped. We infer, then, that his work was original.

Third, his account of the problem is written with the clarity and flair so characteristic of Euler as expositor. It makes his version stand out among the crowd.

Finally, it seems somehow just to discuss Euler's work on a subject first done by others, because there are dozens of instances in mathematics in which *others* are credited with results first discovered by Euler. A case could be made, for instance, that Fourier series, Bessel functions, and even Venn diagrams are all misnamed and should instead be called "Euler series," "Euler functions," and "Euler diagrams." If turnabout is fair play, then perhaps we can discuss Euler's solution to the problem of coincidences without doing irreparable harm.

In any case, Euler posed the problem as follows:

> Given any series of *n* letters *a*, *b*, *c*, *d*, *e*, etc., to find how many ways they can be rearranged so that none returns to the position it initially occupied.

For instance, if we number Montmort's cards from 1 to 13, Euler was addressing the number of ways of dealing them *without* a coincidence. As another illustration, we imagine that a mechanic removes all four tires from a car and inspects them. One might ask in how many ways the tires can be reinstalled so that no tire ends up in its original position. Here it is possible to list the cases and count them to find that there are nine such rearrangements. *Far* harder is to answer the same question were the vehicle an 18-wheeled semi-trailer.

Euler pointed out that, if we impose no restriction on the final position of the letters, there would clearly be *n*! arrangements. We recall, for instance, the $3! = 6$ arrangements of the bakery customers *a*, *b*, and *c* from earlier in this chapter. Obviously not all of these satisfy the constraints of the problem, for in some *a* occupies the initial position, or *b* occupies the intermediate spot, or *c* comes last.

As was his custom, Euler began by considering simple examples in the hope of discerning a broader principle. Ever the symbol-lover, he introduced the notation $\prod(n)$ to represent the number of permutations of the *n* letters *a*, *b*, *c*, *d*, ... in which no letter occupies its original position. (Such a permutation is now called a **derangement**.) He then investigated cases.

If $n = 1$, we have the single letter *a*, whose only permutation keeps it right where it is. Thus $\prod(1) = 0$.

If $n = 2$, we begin with $a\,b$. There are two permutations here—$a\,b$ and $b\,a$—only the latter of which moves each letter to a different spot. So $\prod(2) = 1$.

If $n = 3$, we consider permutations of $a\,b\,c$. As one can see from the bakery queues, only two of the six permutations are derangements, namely $b\,c\,a$ and $c\,a\,b$. Thus $\prod(3) = 2$.

For $n = 4$, there are $4! = 24$ permutations of $a\,b\,c\,d$, but (as was noted in the example of the tires above) only nine move each letter to a different spot. These derangements are underlined in boldface among the 24 permutations below:

$a\,b\,c\,d$	$b\,a\,c\,d$	$c\,a\,b\,d$	***d a b c***
$a\,b\,d\,c$	***b a d c***	***c a d b***	$d\,a\,c\,b$
$a\,c\,b\,d$	$b\,c\,a\,d$	$c\,b\,a\,d$	$d\,b\,a\,c$
$a\,c\,d\,b$	***b c d a***	$c\,b\,d\,a$	$d\,b\,c\,a$
$a\,d\,b\,c$	***b d a c***	***c d a b***	***d c a b***
$a\,d\,c\,b$	$b\,d\,c\,a$	***c d b a***	***d c b a***

Adventurous readers might consider the $5! = 120$ permutations of $a\,b\,c\,d\,e$ and verify that 44 leave no letter in its original position.

Therefore, Euler's initial investigation led him to the sequence:

$$\prod(1) = 0 \quad \prod(2) = 1 \quad \prod(3) = 2 \quad \prod(4) = 9 \quad \text{and} \quad \prod(5) = 44.$$

From this short list, who would be ready to predict $\prod(6)$? And how would anyone find $\prod(12)$? Theoretically, one could list the $12! = 479{,}001{,}600$ permutations of the letters $a\,b\,c\,d\,e\,f\,g\,h\,i\,j\,k\,l$ and then determine which among those half-billion arrangements move all 12 letters around, but it is impossible to imagine anything less fulfilling.

To have any hope of solving this problem, one needs a rule governing $\prod(n)$. Not only did Euler spot the pattern, but he gave a proof—involving a recursive relationship—of why it works. It is an argument both elementary and ingenious.[7]

Theorem. *For $n \geq 3$,*

$$\prod(n) = (n-1)\left[\prod(n-1) + \prod(n-2)\right]. \tag{8.1}$$

---

[7] Euler, *Opera Omnia*, Ser. 1, Vol. 7, pp. 436–438.

Proof. Euler began with n letters $abcde\ldots$. In rearranging these so that no letter returns to its original spot, there are $n-1$ choices for the first letter, since it cannot be a. It made no difference which of the other $n-1$ letters occupied the first slot, for a parallel argument handles each case. For the sake of simplicity, Euler assumed the first letter in his new arrangement is b, and then "whatever number of variations one finds when the letter b occupies the first position, this will be multiplied by $n-1$ to find... $\prod(n)$." The reader will note the multiplication principle at work here.

So, the new permutation begins with b. This led Euler to consider a pair of cases—one in which the letter a stands in the second position and the other in which it does not.

Case 1: Under this scenario, the sequence begins $ba\ldots$. We are then obliged to rearrange the remaining $n-2$ letters $cde\ldots$ so that none returns to its original position. But this is the same problem with which we began, albeit "reduced by two." In terms of the notation above, there are $\prod(n-2)$ different ways to accomplish this.

Case 2: Here we stipulate that the first letter is b and the second is *not* a. The challenge is to rearrange the letters $acde\ldots$ into the $n-1$ slots to the right of b so that c does not return to the third position, nor d to the fourth, nor e to the fifth. Moreover, it must also be the case that a does not fall into the second place. The reason for this last statement, of course, is that we are now in Case 2, where a cannot follow b. If a were to fall into the second slot, the arrangement was already counted in Case 1.

So, ignoring the initial b, we see that the number of permutations for this case is just the number of rearrangements of the $n-1$ letters $acdef\ldots$ in which *none* returns to its original position. This we have denoted by $\prod(n-1)$.

Thus, by Cases 1 and 2, there are $\prod(n-1)+\prod(n-2)$ derangements that begin with b. And because we actually have $n-1$ possibilities for that first letter, the total number of permutations of n letters leaving none of them fixed is

$$\prod(n) = (n-1)\left[\prod(n-1)+\prod(n-2)\right]. \qquad \text{Q.E.D.}$$

This is Euler's recursive relationship. Admittedly, it does not give an explicit, stand-alone formula for $\prod(n)$. Rather, it relates $\prod(n)$ to known values of $\prod(k)$ for $k<n$. This may not be ideal, but it is better than listing and counting.

Starting with $\prod(1) = 0$ and $\prod(2) = 1$, Euler's formula gives:

$$\prod(3) = 2\left[\prod(2) + \prod(1)\right] = 2[1 + 0] = 2,$$

$$\prod(4) = 3\left[\prod(3) + \prod(2)\right] = 3[2 + 1] = 9,$$

$$\prod(5) = 4\left[\prod(4) + \prod(3)\right] = 4[9 + 2] = 44,$$

and

$$\prod(6) = 5\left[\prod(5) + \prod(4)\right] = 5[44 + 9] = 265.$$

Pushing this a bit, we calculate the number of derangements of a dozen letters to be:

$$\prod(12) = 11\left[\prod(11) + \prod(10)\right]$$
$$= 11[14{,}684{,}570 + 1{,}334{,}961] = 176{,}214{,}841.$$

Euler noticed something more, which he called an "excellent relationship."[8] He observed that

$$\prod(2) = 1 = 2 \cdot 0 + 1 = 2\prod(1) + 1,$$

$$\prod(3) = 2 = 3 \cdot 1 - 1 = 3\prod(2) - 1,$$

$$\prod(4) = 9 = 4 \cdot 2 + 1 = 4\prod(3) + 1,$$

$$\prod(5) = 44 = 5 \cdot 9 - 1 = 5\prod(4) - 1,$$

$$\prod(6) = 265 = 6 \cdot 44 + 1 = 6\prod(5) + 1,$$

and in general

$$\prod(n) = n\prod(n - 1) + (-1)^n \text{ for } n \geq 2. \tag{8.2}$$

This recursive formula may be regarded as somewhat preferable to (8.1) because, unlike its predecessor, we now calculate $\prod(n)$ from only *one* prior value of $\prod$, not two.

"It seems miraculous," Euler remarked, that both laws give the same result. But from the second he easily deduced the first by observing that $\prod(n) =$

---

[8]Ibid., pp. 439–440.

$n \prod(n - 1) + (-1)^n$ and $\prod(n - 1) = (n - 1) \prod(n - 2) + (-1)^{n-1}$. Adding these gave

$$\prod(n) + \prod(n - 1) = n \prod(n - 1) + (-1)^n + (n - 1) \prod(n - 2) + (-1)^{n-1}$$
$$= n \prod(n - 1) + (n - 1) \prod(n - 2)$$

because, as Euler graphically put it, $(-1)^n$ and $(-1)^{n-1}$ "destroy one another." A simple transposition led to

$$\prod(n) = (n - 1) \left[\prod(n - 1) + \prod(n - 2) \right],$$

so he had returned to the earlier version of the rule.

It is also possible to push the mathematics in the other direction. That is, let $n = r$ in (8.1) to get $\prod(r) = (r - 1)[\prod(r - 1) + \prod(r - 2)]$, which can be rearranged to

$$\prod(r) - r \prod(r - 1) = - \left[\prod(r - 1) - (r - 1) \prod(r - 2) \right].$$

Apply this repeatedly with $r = n - 1, r = n - 2, \cdots$, down to $r = 3$ to get:

$$\prod(n) - n \prod(n - 1) = - \left[\prod(n - 1) - (n - 1) \prod(n - 2) \right]$$
$$= - \left[- \left[\prod(n - 2) - (n - 2) \prod(n - 3) \right] \right]$$
$$= (-1)^2 \left[\prod(n - 2) - (n - 2) \prod(n - 3) \right]$$
$$= (-1)^3 \left[\prod(n - 3) - (n - 3) \prod(n - 4) \right] = \cdots$$
$$= (-1)^{n-2} \left[\prod(2) - 2 \cdot \prod(1) \right]$$
$$= (-1)^{n-2}[1 - 2 \cdot 0] = (-1)^n.$$

Therefore, $\prod(n) = n \prod(n - 1) + (-1)^n$, which is the recursion of (8.2).

In this way, Euler modified the expression for $\prod(n)$ from a double recursion involving both $\prod(n - 1)$ and $\prod(n - 2)$ to a single recursion in terms of just $\prod(n - 1)$. But he noticed something even more amazing about derangements: that one can find an *explicit* formula for $\prod(n)$ in terms of n alone—a formula allowing direct calculation of $\prod(12)$ without first having to find $\prod(1), \prod(2), \ldots, \prod(10)$, and/or $\prod(11)$. We shall examine this formula, and its unexpected consequences, in the epilogue.

Clearly, Euler had a knack for counting things. The proof just examined is a harbinger of the recursive techniques that, centuries later, would become a hallmark of combinatorics. But, interesting as it is, this theorem seems like a modest first step compared to what we shall see next: Euler's discoveries about the partitioning of numbers.

We begin the story by defining a **partition** of a whole number n to be a representation of n as the sum of other whole numbers. For instance, the number 6 can be written as

$$6, \ 5+1, \ 4+2, \ 4+1+1, \ 3+3, \ 3+2+1, \ 3+1+1+1, \ 2+2+2,$$

$$2+2+1+1, \ 2+1+1+1+1, \quad \text{and} \quad 1+1+1+1+1+1,$$

for a total of 11 partitions. We do not include both $4 + 2$ and $2 + 4$, for these commuted sums are regarded as identical. On the other hand, we do permit the number 6 alone as a "unitary" partition.

Further subclassifications proved important. Euler considered, for instance, the partitions into distinct (i.e., non-repeated) summands. For 6 , there are four of these:

$$6, \quad 5+1, \quad 4+2, \quad 3+2+1.$$

Similarly, he considered those partitions using only odd (but not necessarily distinct) summands. Again, for the case above, there are four:

$$5+1, \quad 3+3, \quad 3+1+1+1, \quad 1+1+1+1+1+1.$$

As a second example, we list below the thirty ways to partition the number 9. Underlined are those which consist of distinct summands, and printed in bold italics are those which consist of *odd summands*. The reader should confirm that, as with the case of 6 above, these kinds of partitions are equally abundant. It was a point not lost on Euler.

9, 8 + 1, 7 + 2, *7 + 1 + 1*, 6 + 3, 6 + 2 + 1, 6 + 1 + 1 + 1,
5 + 4, *5 + 3 + 1*, 5 + 2 + 2, 5 + 2 + 1 + 1, *5 + 1 + 1 + 1 + 1*,
4 + 4 + 1, 4 + 3 + 2, 4 + 3 + 1 + 1, 4 + 2 + 2 + 1,
4 + 2 + 1 + 1 + 1, 4 + 1 + 1 + 1 + 1 + 1, *3 + 3 + 3*,
3 + 3 + 2 + 1, *3 + 3 + 1 + 1 + 1*, 3 + 2 + 2 + 2, 3 + 2 + 2 + 1 + 1,
3 + 2 + 1 + 1 + 1 + 1, *3 + 1 + 1 + 1 + 1 + 1 + 1*, 2 + 2 + 2 + 2 + 1,
2 + 2 + 2 + 1 + 1 + 1, 2 + 2 + 1 + 1 + 1 + 1 + 1,
2 + 1 + 1 + 1 + 1 + 1 + 1 + 1,
1 + 1 + 1 + 1 + 1 + 1 + 1 + 1 + 1

In like manner, we find that there are ten ways to write 10 as the sum of distinct summands and ten ways to write 10 as the sum of odd summands. Each experiment reinforces the phenomenon—there seem to be exactly as many ways to decompose a whole number into distinct pieces as there are to decompose it into (not necessarily distinct) odd ones.

Unfortunately, case by case experimentation soon becomes overwhelming. Anyone following the developments in this chapter may sense that, even for fairly small values of n, there will be a veritable deluge of partitions. The relatively tiny number 24, for instance, boasts 1575 different ones. This is a typical combinatorial phenomenon: a problem that sounds simple but is in fact exceedingly difficult.

The matter of partitions was brought to Euler's attention in a 1740 letter from Philippe Naudé.[9] The latter had written to inquire about Euler's recent triumph in summing $\sum_{k=1}^{\infty} 1/k^2$, but he also asked what Euler knew about the number of different ways a positive integer m could be expressed as the sum of n distinct whole number summands. And what, he asked, if the summands need not be distinct?

Euler apparently had never considered a problem like this before but was more than up to the task. He responded to Naudé within weeks, attributing his tardiness (!) to troubles with deteriorating vision. More remarkable even than the promptness of his reply is the mathematical insight he employed in crafting it. A thorough exposition of Euler's ideas can be found in Book I, Chapter 16, of his *Introductio*. We shall here address only the proof establishing an equal number of partitions into distinct and into odd summands.

In mulling over the problem, Euler somehow recognized a link between counting partitions and multiplying algebraic binomials. He expressed it in this fashion:

> If the product $Q = (1+x)(1+x^2)(1+x^3)(1+x^4)(1+x^5)(1+x^6) \cdots$ is developed, we have the series
>
> $$1 + x + x^2 + 2x^3 + 2x^4 + 3x^5 + 4x^6 + 5x^7 + 6x^8 + 8x^9 + 10x^{10} + \cdots$$
>
> in which each coefficient indicates the number of different ways in which the exponent can be expressed as the sum of different numbers.

This assertion requires some thought. Yet, once understood, it seems self-evident.

---

[9] Weil, pp. 276–277.

Consider, for instance, the coefficient of x^6 in the expansion of Q. Clearly x^6 occurs in the product

$$(1 + x)(1 + x^2)(1 + x^3)(1 + x^4)(1 + x^5)(1 + x^6) \cdots$$

when x^6 in the sixth factor multiplies all the other 1s. It occurs again when the x^5 multiplies x, giving $x^{5+1} = x^6$. A third x^6 comes from $x^4 \cdot x^2 = x^{4+2} = x^6$, and the last is generated by the three-way product $x^3 \cdot x^2 \cdot x = x^{3+2+1} = x^6$.

So there are exactly as many x^6 terms in the expansion of Q as there are ways of decomposing 6 into the sum of *different* whole numbers. The reason we say "different," of course, is because no repetition occurs among the factors in the product. Thus x^6 occurs four times because 6 has the four partitions into distinct summands: $6, 5 + 1, 4 + 2, 3 + 2 + 1$.

Similarly, the term $8x^9$ in the expansion of Q arises as the sum of

$$x^9, \quad x^{8+1}, \quad x^{7+2}, \quad x^{6+3}, \quad x^{5+4}, \quad x^{6+2+1}, \quad x^{5+3+1}, \quad \text{and} \quad x^{4+3+2}.$$

These exponents are precisely the eight underlined partitions in the display above.

Euler had perceived this connection between partitions of integers and algebraic expansions. But he had another trick up his sleeve. He introduced

$$R = \frac{1}{(1 - x)(1 - x^3)(1 - x^5)(1 - x^7) \cdots},$$

the *reciprocal* of an infinite product. What possible bearing could this have on the matter?

The link became apparent when Euler invoked the summation formula for an infinite geometric series:

$$\frac{1}{1 - a} = 1 + a + a^2 + a^3 + a^4 + \cdots.$$

Repeatedly using this summation with $a = x$, $a = x^3$, $a = x^5$, etc., he concluded that

$$R = \frac{1}{1 - x} \times \frac{1}{1 - x^3} \times \frac{1}{1 - x^5} \times \frac{1}{1 - x^7} \times \cdots$$

$$= (1 + x + x^2 + x^3 + \cdots)(1 + x^3 + x^6 + x^9 + \cdots)(1 + x^5 + x^{10} + \cdots)$$

$$\times (1 + x^7 + x^{14} + \cdots) \cdots,$$

which is not merely an infinite product but an infinite product of infinite series!

What makes it germane to the problem at hand is that each exponent can be regarded as the sum of odd numbers. That is, Euler rewrote the product as:

$$R = (1 + x^1 + x^{1+1} + x^{1+1+1} + \cdots)(1 + x^3 + x^{3+3} + x^{3+3+3} + \cdots)$$
$$\times(1 + x^5 + x^{5+5} + \cdots)\cdots. \tag{8.3}$$

Upon multiplying and combining like terms, we get

$$R = 1 + x + x^2 + 2x^3 + 2x^4 + 3x^5 + 4x^6 + 5x^7 + 6x^8 + 8x^9 + 10x^{10} + \cdots.$$

Again we ask, "Why do we have four x^6 terms in this expansion?" Clearly, from the form of the product in (8.3), it is because there are four ways to represent 6 as a sum of odd numbers, these being the only exponents available. That is, we get x^{5+1}, x^{3+3}, $x^{3+1+1+1}$, $x^{1+1+1+1+1+1}$. These exponents are precisely the partitions of 6 into odd components, as noted earlier. In a similar fashion, there are eight x^9 terms because 9 arises as a sum of odd numbers in the eight ways denoted in boldface italics above.

The reader should compare the first few terms of the series for R with the corresponding terms of the series for Q. Their perfect agreement must have thrown Euler into a state of euphoria. It could not have been an accident.

To summarize, Euler had observed that the number of ways of writing n as the sum of distinct whole numbers is the coefficient of x^n in the expansion of

$$Q = (1 + x)(1 + x^2)(1 + x^3)(1 + x^4)(1 + x^5)\cdots,$$

whereas the number of ways of writing n as the sum of (not necessarily distinct) odd whole numbers is the coefficient of x^n in the expansion of

$$R = \frac{1}{(1 - x)(1 - x^3)(1 - x^5)(1 - x^7)\cdots}.$$

The stage was set for one of Euler's most ingenious deductions. In his words,[10]

Theorem. *The number of different ways a given number can be expressed as the sum of different whole numbers is the same as the number of ways in which that same number can be expressed as the sum of odd numbers, whether the same or different.*

---

[10]Euler, *Introduction to Analysis of the Infinite*, Book I, pp. 275–276.

Proof. Again letting $Q = (1 + x)(1 + x^2)(1 + x^3)(1 + x^4)(1 + x^5) \cdots$, Euler introduced $P = (1 - x)(1 - x^2)(1 - x^3)(1 - x^4)(1 - x^5) \cdots$ so that

$$PQ = (1 - x)(1 + x)(1 - x^2)(1 + x^2)(1 - x^3)(1 + x^3) \cdots$$
$$= (1 - x^2)(1 - x^4)(1 - x^6)(1 - x^8) \cdots.$$

Because "all of the factors of PQ are contained in P," he concluded that

$$\frac{1}{Q} = \frac{P}{PQ} = (1 - x)(1 - x^3)(1 - x^5)(1 - x^7) \cdots,$$

and so

$$Q = \frac{1}{(1 - x)(1 - x^3)(1 - x^5)(1 - x^7) \cdots} = R.$$

In short, Q and R are the same thing. Consequently, the coefficient of x^n in the expansion of Q must equal the coefficient of x^n in the expansion of R. But, as noted above, these (equal) coefficients are, respectively, the number of partitions of n into distinct summands and the number of partitions of n into odd summands. The proof is complete.   Q.E.D.

It is hard to decide which is more remarkable: the equality described by the theorem or Euler's proof of it. The latter is simple. It is elegant. And it is comprehensive in establishing the results for all n with a single argument.

In literature or art or theatre, one occasionally encounters a work so novel, so powerful as to take one's breath away. In mathematics, such reactions are harder to come by, for our discipline appeals directly to the rational, not the emotional, faculties. Euler's proof, delivered almost by return mail to the inquisitive Naudé, is about as close to breathtaking as mathematics is likely to get.

Epilogue

Because this chapter has focused on two particular combinatorial results, so will the epilogue. Reversing the order, we shall first say a word about partitions and then return to the intriguing problem of derangements.

Euler's research on the partitioning of numbers was not limited to the theorem above. As noted, Chapter XVI of Book I of his *Introductio* contains a thorough development of these ideas. Having consumed many pages on such matters, he wrote, "There remain a few problems of this type which are worth notice, and which also have some use for the understanding of the nature of numbers." Rather like a child with a new toy, he seemed to be having too much fun to stop.

As a case in point, Euler introduced the infinite product

$$P(x) = (1 + x)(1 + x^2)(1 + x^4)(1 + x^8)(1 + x^{16}) \cdots,$$

"in which each exponent is the double of its predecessor." Reasoning as before, Euler knew that, upon expanding this product, he would get as many x^n terms as there were ways (if any) of writing n as the sum of powers of 2. To determine the infinite product exactly, he employed the following strategy.

Assume that, after multiplying, one finds

$$(1 + x)(1 + x^2)(1 + x^4)(1 + x^8)(1 + x^{16}) \cdots = P(x)$$

$$= 1 + \alpha x + \beta x^2 + \gamma x^3 + \delta x^4 + \epsilon x^5 + \cdots \qquad (8.4)$$

where the coefficients $\alpha, \beta, \gamma, \delta, \cdots$ are yet to be determined. Euler observed that

$$\frac{P(x)}{1 + x} = (1 + x^2)(1 + x^4)(1 + x^8)(1 + x^{16}) \cdots.$$

Here the right-hand side is obviously $P(x^2)$, and so:

$$\frac{P(x)}{1 + x} = P(x^2) = 1 + \alpha x^2 + \beta x^4 + \gamma x^6 + \delta x^8 + \cdots.$$

Cross-multiplication yields:

$$P(x) = (1 + x)(1 + \alpha x^2 + \beta x^4 + \gamma x^6 + \cdots)$$

$$= 1 + x + \alpha x^2 + \alpha x^3 + \beta x^4 + \beta x^5 + \gamma x^6 + \gamma x^7 + \cdots.$$

Comparing this to the original expression for $P(x)$ in (8.4), Euler deduced that $\alpha = 1, \beta = \alpha, \gamma = \alpha, \delta = \beta, \epsilon = \beta$, and so on. Therefore, all of the original coefficients are equal to 1, and so:

$$(1 + x)(1 + x^2)(1 + x^4)(1 + x^8)(1 + x^{16}) \cdots = 1 + x + x^2 + x^3 + x^4 + x^5 + \cdots.$$

Euler emphasized, "Since each whole number occurs exactly once as an exponent in the series... it follows that every whole number can be expressed as the sum of terms in the geometric progression ... 1, 2, 4, 8, 16, 32, ... *and this sum is unique*" [italics added].[11] He had thereby established the unique binary representation of each integer.

---

[11] Ibid., pp. 277–278.

Worth noting here is not so much the final result, which long predated Euler. Rather, it is the unusual route by which he obtained it: employing the infinite products that had proved so useful in his analysis of partitions. This demonstrates yet again a mind able to examine a familiar problem under a penetrating new light.

As promised, the other topic of this epilogue is a return to Euler's recursive expression for the number of derangements of n distinguishable objects. We have seen Euler go from the "double" recursive formula

$$\prod(n) = (n-1)\left[\prod(n-1) + \prod(n-2)\right]$$

to the simpler recursion $\prod(n) = n\prod(n-1) + (-1)^n$. But Euler pushed further, to a closed-form expression for $\prod(n)$ requiring no prior calculation of any $\prod(k)$. Better yet, the result has some profound implications.[12]

Theorem. *For all $n \geq 1$,*

$$\prod(n) = n!\left[1 - \frac{1}{1!} + \frac{1}{2!} - \frac{1}{3!} + \frac{1}{4!} - \cdots + \frac{(-1)^n}{n!}\right].$$

Proof. This is a straightforward induction. If $n = 1$, $\prod(1) = 0$ as is $1![1 - \frac{1}{1!}]$. Next, assume the result holds for $n = k$. By Euler's second recursion,

$$\prod(k+1) = (k+1)\prod(k) + (-1)^{k+1}$$

$$= (k+1)\left(k!\left[1 - \frac{1}{1!} + \frac{1}{2!} - \frac{1}{3!} + \frac{1}{4!} - \cdots + \frac{(-1)^k}{k!}\right]\right)$$

$$+ (-1)^{k+1}$$

$$= (k+1)!\left[1 - \frac{1}{1!} + \frac{1}{2!} - \frac{1}{3!} + \frac{1}{4!} - \cdots\right.$$

$$\left. + \frac{(-1)^k}{k!} + \frac{(-1)^{k+1}}{(k+1)!}\right],$$

and the theorem follows.   Q.E.D.

With this, we see immediately that the number of derangements of six items is

[12] Euler, *Opera Omnia*, Ser. 1, Vol. 7, pp. 542–545.

$$\prod(6) = 6! \left[1 - \frac{1}{1!} + \frac{1}{2!} - \frac{1}{3!} + \frac{1}{4!} - \frac{1}{5!} + \frac{1}{6!} \right]$$
$$= 720 - 720 + 360 - 120 + 30 - 6 + 1 = 265,$$

exactly as we found above. In a similar fashion, we calculate directly—without need of recursion—that $\prod(12) = 176,214,841$.

This allows us to answer a related question: if an ordered set of items is randomly permuted, what is the *probability* that none of the items returns to its original position? What, in other words, is the probability of a derangement? For example, if we remove a dozen eggs from a carton, wash each one, and then replace them at random, what is the chance that no egg ends up where it began?

It is clear that n distinguishable items can be permuted in $n!$ different ways, of which $\prod(n)$ are derangements that return none of the items to its starting point. Therefore the probability of no item's being returned to its original position is:

$$p_n = \frac{\prod(n)}{n!} = \frac{n! \left[1 - \frac{1}{1!} + \frac{1}{2!} - \frac{1}{3!} + \frac{1}{4!} - \cdots + \frac{(-1)^n}{n!} \right]}{n!}$$
$$= 1 - \frac{1}{1!} + \frac{1}{2!} - \frac{1}{3!} + \frac{1}{4!} - \cdots + \frac{(-1)^n}{n!}.$$

For small values of n, we tabulate the probabilities

n	$\prod(n)$	Probability $= p_n = \prod(n)/n!$
1	0	$p_1 = 0$
2	1	$p_2 = \frac{1}{2!} = 0.5$
3	2	$p_3 = \frac{2}{3!} \approx 0.333333$
4	9	$p_4 = \frac{9}{4!} = 0.375000$
5	44	$p_5 = \frac{44}{5!} \approx 0.366667$
6	265	$p_6 = \frac{265}{6!} \approx 0.368056$

In our example, the chance that no egg returns to its original spot in the carton is

$$p_{12} = \frac{\prod(12)}{12!} = \frac{176214841}{479001600} \approx 0.3678794413.$$

We observe that this answer is very close to p_6. Moreover, if the carton held *two* dozen eggs, the probability of none returning to its starting point is

$$p_{24} = \frac{\prod(24)}{24!} \approx 0.3678794412,$$

which is extremely close to the "one dozen" answer.

This stabilization of probabilities is no accident, as Euler explained in a 1751 paper.[13] As n grows without bound, we have

$$\lim_{n \to \infty} p_n = \lim_{n \to \infty} \left[1 - \frac{1}{1!} + \frac{1}{2!} - \frac{1}{3!} + \frac{1}{4!} - \cdots + \frac{(-1)^n}{n!} \right]$$

$$= e^{-1} = 1/e \approx 0.3678794412,$$

where we revisit Euler's series expansion of e^x developed in Chapter 2.

Note also that the rate of convergence is extremely rapid. Because we are summing an alternating series, we know its nth partial sum differs from its limit $1/e$ by less than $1/(n+1)!$. Thus the likelihood of a derangement very quickly stabilizes around 0.36788. Euler himself observed that the value is essentially unchanged for all $n \geq 20$, which means that the probability of no egg's returning to its original position is the same whether we begin with two dozen eggs or *two billion* dozen eggs.[14] And this probability, strange to say, is almost exactly $1/e$.

It seems miraculous that e should appear, almost out of nowhere, in this combinatorial problem. There is certainly no obvious link between eggs moving around a carton and the base of the natural logarithms, but students of calculus will recall that the number e is ubiquitous in mathematical applications. It pops up in some of the strangest places—and none more so than this.

Volumes could be written about advances in combinatorial theory over the centuries since Euler considered these ideas.[15] Suffice it to say that combinatorics, a subject that exploded in the twentieth century, has roots running back into the mathematical past. And at least a few of these can be traced to Leonhard Euler, who seems never to have met a math problem he didn't like. In the field of combinatorics, as in the fields of number theory and analysis and geometry, he left footprints deep and lasting.

---

[13] Ibid., pp. 11–25.

[14] Ibid., p. 25.

[15] A very nice survey is H. L. Alder's "Partition Identities—from Euler to the Present," *The American Mathematical Monthly*, Vol. 76, No. 7, 1969, pp. 733–746.

Conclusion

Surveying once again Euler's *Opera Omnia* on those sagging library shelves, I recall the words of André Weil:

> No mathematician ever attained such a position of undisputed leadership in all branches of mathematics, pure and applied, as Euler did for the best part of the eighteenth century.[1]

These eight chapters, I hope, have provided ample support for Weil's assertion. Collectively, I have discussed three dozen of Euler's original proofs drawn from thirteen different volumes of the *Opera Omnia*. They demonstrate his ability to address questions old and new, discrete and continuous, algebraic and analytic—and in the process to carry mathematics to places undreamed of.

Yet my book completely ignores vast portions of Euler's work. I made no mention of his contributions to differential equations or to the calculus of variations. I omitted his solution to the Königsberg Bridge problem, which gave birth to modern graph theory; his development of the gamma function, which cleverly extended the factorial to non-integer values; and his paper on the so-called Euler–Descartes formula (i.e., $V + F = E + 2$), which played the central role in early combinatorial topology.

Perhaps even more glaring is my omission of applications. The *Opera Omnia* contains nearly forty *volumes* of what we now call "applied mathematics," ranging from mechanics to optics, from music to naval science. This could be fertile ground for a string of sequels.

For now, I shall end with a word or two about Euler's legacy.

As observed throughout this book, his mathematics did not always display the rigor and precision of today's, most particularly in his cavalier use of the infinite. These shortcomings have given ammunition to mathematicians who

---

[1] Weil, p. 169.

criticize his work as primitive, intuitive, and decidedly pre-modern. They have a point.

On the other hand, one could reasonably ask whether modern mathematics would even *exist* without him. It is true that he sometimes proceeded heuristically, relying on intuition as much as logic. But had he been dissuaded by a standard of rigor beyond his reach, would Euler have forgone his remarkable journey? One is reminded of Horace Lamb's perceptive remark:

> A traveler who refuses to pass over a bridge until he has personally tested the soundness of every part of it is not likely to go far; something must be risked, even in mathematics.[2]

Euler took his share of risks, but for this I believe the mathematical world should be ever grateful.

---

[2] Kline, p. 468.

In this spirit, I cite a passage from Galileo's *Dialogues Concerning Two New Sciences* in which he described the promise and peril of innovation. Although written in an earlier century, Galileo's words relate perfectly to Euler's achievements.

"My regard for the inventor of the harp," he wrote,

> is not made less by knowing that the instrument was very crudely constructed and still more crudely played. Rather, I admire [the inventor] more than I do the hundreds of craftsmen who in ensuing centuries have brought this art to the highest perfection....

And Galileo concluded with this striking observation:

> To apply oneself to great inventions, starting from the smallest beginnings, is no task for ordinary minds.[3]

Leonhard Euler was an inventor, an explorer, an artist. With an enthusiasm that resonates after two centuries and more, he ventured into parts unknown—not outward to the physical world but inward to that domain where, as Bertrand Russell once put it, "pure thought can dwell as in its natural home."[4] As with any great explorer, he now and then took a wrong turn or missed an important landmark. Nonetheless, like Galileo's ancient harpist, Euler deserves our utmost admiration. Laboring in the semi-darkness, working with only the power of his unmatched imagination, he journeyed to the mathematical frontier and beyond.

I hope, over these eight chapters, the case has been made:

It was no ordinary mind that gave us this extraordinary mathematics.

---

[3] Stillman Drake, trans., *Discoveries and Opinions of Galileo*, Doubleday Anchor Books, Garden City, NY, 1957, p. 1.

[4] Robert E. Egner and Lester E. Dennon, eds., *The Basic Writings of Bertrand Russell: 1903–1959*, Simon & Schuster, New York, 1961, p. 254.

Euler's *Opera Omnia*

The greatest resource for anyone interested in the mathematics of Leonhard Euler is his collected works, the *Opera Omnia*. Their publication, begun in 1911, has consumed the remainder of the twentieth century. To date, six dozen volumes (more or less) have appeared, but the Eulerian gusher has not yet run dry.

A number of challenges presented themselves. One, as should be expected for a venture of this magnitude, was the matter of funding. Another was the intervening devastation of two World Wars. Particularly vexing, strange to say, was the "simple" job of identifying everything that Euler wrote. At the time of his death, 560 items had appeared in print, but the St. Petersburg Academy continued posthumous publication for decades thereafter. A census from 1843 contained 756 items.[1] Then, just when everyone thought the backlog had been eradicated, another 61 hitherto unknown manuscripts were discovered. By the time Gustaf Eneström (1852–1923) completed a survey of Euler's works in the early twentieth century, their number had grown to 866. Eneström's list was used to organize the *Opera Omnia*.[2]

In 1909, the Swiss Academy of Sciences voted to publish the collection— whatever its size—and chose Ferdinand Rudio (1856–1929) to oversee the task. This proved an excellent choice, for Rudio was an Euler enthusiast of the first order. He tirelessly threw himself into the project, one whose completion he knew he would not live to see.

---

[1] For the actual list, see Fuss, Vol. 1, pp. LVII-CXX.

[2] A more detailed story of the collection and publication of Euler's work can be found in S. B. Engelsman's "What You Should Know about Euler's *Opera Omnia*," *Nieuw Archief voor Wiskunde*, 4th Series, Vol. 8, No. 1, 1990, pp. 67–79.

A typical volume of the *Opera Omnia* is large, running from 400 to 500 pages—although some contain over 700. In size and weight, such a volume resembles its counterpart from (say) the *Encyclopedia Britannica*. No one short of an athlete could carry more than five or six at once, and to cart off the entire collection—over 25,000 pages in all—would require a forklift.

In terms of organization, the *Opera Omnia* is divided into four major parts, or "series," each of which contains multiple volumes. Euler's pure mathematics—algebra, analysis, number theory, etc.—is found in the 29 volumes of Series I. His mechanics, engineering, and astronomy will, when completed, occupy the 31 volumes of Series II. Series III will contain 12 volumes on physics and "miscellanea." And it is anticipated that the relatively new Series IV will contain 8 volumes of Euler's scientific correspondence (in Part IVA) and 7 volumes of other manuscripts (in Part IVB). In two cases—Series I, Volume 16 and Series II, Volume 11—a single volume is split into a pair of books.

Many volumes open with a summary of the contents, and a few contain relevant papers by other authors. For example, the editors followed Euler's incorrect proof of the fundamental theorem of algebra (see Chapter 6) with Gauss's critique of the same. Likewise, the second part of Series II, Volume 11 is Clifford Truesdell's book (in English) on the early history of the theory of elastic bodies. However, the vast majority of *Opera Omnia* pages contain the words of Euler and Euler alone.

It was decided at the outset that items would be published in the languages in which they originally appeared. Thus, the *Opera Omnia* features mainly Latin and French, with a smattering of German—discouraging news for the English-only reader. Indeed, English translations of Euler's books and papers are rare. As noted below, his algebra textbook, his classic *Introductio in analysin infinitorum*, and his *Letters to a German Princess* have been translated, but these are the exception and not the rule.

Availability is another problem. Because of the nature and expense of the *Opera Omnia*, its volumes will not be found in the neighborhood library or bookshop. Generally one must travel to a major research library to have any hope of seeing the collection in person.

A few papers have been translated and are readily available. One can consult, for instance, David Smith's *A Source Book in Mathematics*, Dirk Struik's *A Source Book in Mathematics: 1200–1800*, Ronald Calinger's *Classics of Mathematics*, or John Fauvel and Jeremy Gray's *The History of Mathematics: A Reader*—all of which are referenced in the Chapter Notes. A famous paper on continued fractions has been translated by Myra Wyman (a Latin scholar) and her son Bostwick (a mathematician).[3]

Another resource for the Latin/French/German-impaired is the November 1983 issue of *Mathematics Magazine*. Commemorating the bicentennial of Euler's death, it features a collection of articles (albeit not direct translations) about his work as well as an extensive glossary of mathematical terms bearing his name.

With these preliminaries aside, we now give a quick overview of the *Opera Omnia*. Here we adopt the convention of writing "Vol. II.5" to indicate the fifth volume of the second series. Within each series, works are grouped by subject, and within each subject, by date of publication. Thus, for instance, Euler's geometry appears in Volumes I.26—I.29, with the earliest papers in Vol. I.26 and posthumous publications in Vol. I.29. If Euler wrote a comprehensive text on the subject, this precedes the collected papers regardless of publication date.

The list that follows indicates each *Opera Omnia* volume, its date of publication, its page count, and its subject matter.

Euler's *Opera Omnia*

Series I—Pure Mathematics

Vol. I.1 (published 1911/ 651 pp.)—the 1770 text on algebra
  Available in English as *Elements of Algebra*, trans. John Hewlett, Springer-Verlag, New York (1840 Reprint)
Vol. I.2 (published 1915/ 611 pp.)—papers on number theory
Vol. I.3 (published 1917/ 543 pp.)—papers on number theory
Vol. I.4 (published 1941/ 431 pp.)—papers on number theory
Vol. I.5 (published 1944/ 370 pp.)—papers on number theory
Vol. I.6 (published 1921/ 509 pp.)—papers on the theory of equations
Vol. I.7 (published 1923/ 577 pp.)—papers on combinatorics and probability
Vol. I.8 (published 1922/ 390 pp.) ⎫ the 1748 *Introductio in analysin*
Vol. I.9 (published 1945/ 402 pp.) ⎭ *infinitorum* (2 Vols.)
  Available in English as *Introduction to Analysis of the Infinite*, trans. John Blanton, Springer-Verlag, New York, 1988.
Vol. I.10 (published 1913/ 676 pp.)—the 1755 text on differential calculus

---

[3]B.F. Wyman and M.F. Wyman (trans.) in *Mathematical Systems Theory*, No. 18, Vol. 4, 1985, pp. 295–328.

Vol. I.11 (published 1913/ 462 pp.) ⎫
Vol. I.12 (published 1914/ 542 pp.) ⎬ —the 1768 text on integral calculus (3 Vols.)
Vol. I.13 (published 1914/ 505 pp.) ⎭
Vol. I.14 (published 1925/ 617 pp.)—papers on infinite series
Vol. I.15 (published 1927/ 722 pp.)—papers on infinite series
Vol. I.16—part 1 (published 1933/ 355 pp.) ⎫
Vol. I.16—part 2 (published 1935/ 328 pp.) ⎬—papers on infinite series
Vol. I.17 (published 1914/ 457 pp.)—papers on integration
Vol. I.18 (published 1920/ 475 pp.)—papers on integration
Vol. I.19 (published 1932/ 492 pp.)—papers on integration
Vol. I.20 (published 1912/ 370 pp.)—papers on elliptic integrals
Vol. I.21 (published 1913/ 380 pp.)—papers on elliptic integrals
Vol. I.22 (published 1936/ 420 pp.)—papers on differential equations
Vol. I.23 (published 1938/ 455 pp.)—papers on differential equations
Vol. I.24 (published 1952/ 308 pp.)—the 1744 text on the calculus of variations
Vol. I.25 (published 1952/ 342 pp.)—papers on the calculus of variations
Vol. I.26 (published 1953/ 362 pp.)—papers on geometry
Vol. I.27 (published 1954/ 400 pp.)—papers on geometry
Vol. I.28 (published 1955/ 381 pp.)—papers on geometry
Vol. I.29 (published 1956/ 446 pp.)—papers on geometry

Series II—Mechanics and Astronomy

Vol. II.1 (published 1912/ 407 pp.) ⎫
Vol. II.2 (published 1912/ 459 pp.) ⎬—the 1736 text on mechanics (2 Vols.)
Vol. II.3 (published 1948/ 327 pp.) ⎫ the 1765 text on motion of rigid
Vol. II.4 (published 1950/ 358 pp.) ⎬ bodies (2 Vols.)
Vol. II.5 (published 1957/ 324 pp.)—papers on mechanics
Vol. II.6 (published 1957/ 302 pp.)—papers on mechanics
Vol. II.7 (published 1958/ 326 pp.)—papers on mechanics
Vol. II.8 (published 1964/ 417 pp.)—papers on mechanics of rigid bodies
Vol. II.9 (published 1968/ 441 pp.)—papers on mechanics of rigid bodies
Vol. II.10 (published 1947/ 450 pp.)—papers on mechanics of elastic bodies
Vol. II.11—part 1 (published 1957/ 382 pp.)—papers on mechanics of elastic bodies
Vol. II.11—part 2 (published 1960/ 428 pp.)—Clifford Truesdell's *The Rational Mechanics of Flexible or Elastic Bodies, 1638–1788.*
Vol. II.12 (published 1954/ 288 pp.)—papers on fluid mechanics
Vol. II.13 (published 1955/ 374 pp.)—papers on fluid mechanics

Vol. II.14 (published 1922/ 481 pp.)—Euler's translation of Benjamin Robins's book on artillery, with annotations

Vol. II.15 (published 1957/ 318 pp.)—papers on the theory of machines

Vol. II.16 (published 1979/ 327 pp.)—papers on the theory of machines

Vol. II.17 (published 1982/ 312 pp.)—papers on the theory of machines

Vol. II.18 (published 1967/ 427 pp.)⎱ the 1749 text on naval science
Vol. II.19 (published 1972/ 459 pp.)⎰ (2 Vols.)

Vol. II.20 (published 1974/ 275 pp.)—papers on naval science

Vol. II.21 (published 1978/ 241 pp.)—papers on naval science

Vol. II.22 (published 1958/ 411 pp.)—the 1772 text on the theory of lunar motion

Vol. II.23 (published 1969/ 336 pp.)—papers on solar and lunar motion

Vol. II.24 (published 1991/ 326 pp.)—papers on solar and lunar motion

Vol. II.25 (published 1960/ 331 pp.)—papers on the theory of astronomical perturbation

Vol. II.26—to appear

Vol. II.27—to appear

Vol. II.28 (published 1959/ 331 pp.)—papers on motion of planets and comets

Vol. II.29 (published 1961/ 420 pp.)—papers on astronomical precession and nutation

Vol. II.30 (published 1964/ 351 pp.)—papers on eclipses and parallax

Vol. II.31 (published 1996/ 378 pp.)—papers on tides and geophysics

Series III—Physics and Miscellanea

Vol. III.1 (published 1926/ 590 pp.)—papers on general physics and acoustics

Vol. III.2 (published 1942/ 429 pp.)—the 1738 text on basic school arithmetic

Vol. III.3 (published 1911/ 510 pp.)⎱ the 1769 text on optics (2 Vols.)
Vol. III.4 (published 1912/ 543 pp.)⎰

Vol. III.5 (published 1962/ 395 pp.)—papers on optics

Vol. III.6 (published 1962/ 395 pp.)—papers on optics

Vol. III.7 (published 1964/ 247 pp.)—papers on optics

Vol. III.8 (published 1969/ 266 pp.)—papers on optics

Vol. III.9 (published 1973/ 328 pp.)—papers on optics, including E.A. Fellmann's essay on Euler's place in the history of optics (in German)

Vol. III.10—to appear

Vol. III.11 (published 1960/ 312 pp.)⎱ the 1768 *Letters to a German*
Vol. III.12 (published 1960/ 310 pp.)⎰ *Princess* (2 Vols.)

Available in English as *Letters of Euler on Different Subjects in Natural Philosophy, Vols. I and II*, Arno Press, New York, 1975.

Series IVA—Correspondence

Vol. IVA.1 (published 1975/666 pp.)—index of Euler's known correspondence
Vol. IVA.2 (published 1998/747 pp.)—correspondence with Johann and
  Nicolaus Bernoulli
Vol. IVA.3—to appear
Vol. IVA.4—to appear
Vol. IVA.5 (published 1980/611 pp.)—correspondence with Clairaut,
  d'Alembert and Lagrange
Vol. IVA.6 (published 1986/453 pp.)—correspondence with Maupertuis and
  Frederick the Great
Vol. IVA.7—to appear
Vol. IVA.8—to appear

Series IVB—Manuscripts

To appear

Index